I0099776

Ancient Wine and the Bible
The Case for Abstinence

by David R. Brumbelow
Foreword by Paige Patterson

Ancient Wine and the Bible:
The Case for Abstinence

Published by Free Church Press
P.O. Box 1075
Carrollton, Georgia 30112 USA
www.freechurchpress.com

Copyright © 2011 David R. Brumbelow
All Rights Reserved.

Printed in the United States of America.
By Lightning Source Inc.
1246 Heil Quaker Blvd.
La Vergne, TN USA 37086

Except where otherwise indicated, Scripture taken from the New King James Version. Copyright 1979, 1980, 1982 by Thomas Nelson, Inc. Used by Permission. All rights reserved.

ISBN 9780982656129
Library of Congress Control Number   2011936729

Contact author:
David R. Brumbelow
P. O. Box 300
Lake Jackson, TX 77566
gulfcoastpastor.blogspot.com

Permission granted to use brief excerpts if followed by: Ancient Wine and the Bible, David R. Brumbelow, freechurchpress.com

Cover Design: Jessica Anglea www.jessicaanglea.com
Text Design: Debbie Patrick, Vision Run, www.visionrun.com
Managing Editor: Chris Gilliam, Orange Lake, Florida

free
church
press

"Hip Christianity may make the appeal for the use of alcohol as a witnessing tool, but David Brumbelow's exhaustive research presents a powerful case for abstinence. I commend him for taking an unpopular stand against a popular practice."

-Dr. Jim Richards, *Executive Director, Southern Baptists of Texas Convention*

"Brumbelow has done an amazing and splendidly helpful job in analyzing the data about alcohol, presenting the evidence, and applying the proper and obvious conclusions. He explodes all the modern arguments in favor of social drinking, including the myth that the ancients had no way of preserving grape juice. If it triggers a resolve in one young person to never take his/her first drink, it will be worth all the author's trouble in doing the tremendous research involved in making this book. If it causes one soul sinking in the pollution of alcoholism to realize there is salvation, freedom and eternal life in the Lord Jesus Christ, it will be a howling success.If it brings one defeated Christian to a place of hope, helping him or her make a fresh start in the Christian life, it will be a profitable volume. It should do all the above; and more! I happily, enthusiastically, earnestly, wholeheartedly recommend Ancient Wine and the Bible to every pastor, teacher, evangelist, deacon and humble Christian in America. It will be a good investment paying rich dividends in the days ahead."

-Dr. Robert L. Sumner, *Editor, The Biblical Evangelist,*
134 Salisbury CircleLynchburg,Virginia24502-5056,
434/237-0132; biblicalevangelist.org

David R. Brumbelow's passionate plea for abstinence from alcoholic drink needs to be heard attentively in our day. Having lived and served as a missionary in several countries where the lives of wives and children have been devastated by an alcoholic husband and father have taught me that the farther one stays from alcoholic drink the better one will be in providing a good example and in helping those who are enslaved by alcoholism. Brumbelow's admonition should reverberate in our hearts and minds: "We are to be willing to deny ourselves, not abuse our liberties, out of love for our brothers in Christ and out of love for those who are nonbelievers. Don't be a stumbling block to others."

- Dr. Daniel R. Sanchez (*DMin, PhD*), *Professor of Missions,*
Southwestern Baptist Theological Seminary, Fort Worth, Texas
(*swbts.edu*)

"David Brumbelow has left no stone unturned in making the case that the Church of the Lord Jesus Christ should totally abstain from the use of alcoholic beverages. As the daughter of an alcoholic father, I can speak for the millions whose lives have been diminished, damaged and destroyed by the legal drug (alcohol). For those of us who have felt the love, acceptance and safety of a sober church, I want to thank David Brumbelow for writing Ancient Wine and the Bible. This book shines a bright light on the biblical truth that, in God's economy, beverage alcohol has absolutely no benefits and a myriad of deficits. Comprehensive and thoroughly researched, Ancient Wine and the Bible deserves to be read, considered and heeded."

-Mrs. John Hatch. *Alice Hatch is a pastor's wife and mother of three girls. They have ministered in churches in Texas and Washington.*
John Hatch now serves as Director of Missions, Gregg Baptist Association, Longview, Texas

"Pastor David Brumbelow has done the Church of Jesus Christ a great service by penning Ancient Wine and the Bible. In a day in which so-called social drinking, or drinking in moderation, as it has come to be known, is becoming increasingly popular and accepted by professing Christians, including pastors who should know better, this book couldn't have come at a more opportune time. It is no exaggeration to assert that the social drinking craze is creating a tidal wave of tsunami proportions that is beginning to inundate the Church in America, and it is and will continue to cause massive casualties and leave great destruction in its pathway. As a pastor, it grieves my heart to see the people of God being tricked by the Devil into believing that God now condones and encourages His people to imbibe of that which He said in the Old Testament is a mocker. There was a time in the recent past when God's people were in almost unanimous agreement that the Word of God condemns the use of all alcoholic beverages in all forms. Who has changed? Certainly not God! I urge all who take up this book to read it prayerfully and with a desire to understand what the Bible teaches on this serious matter. Then, armed with the truth, stand for righteousness and refuse to fall for the propaganda being spewed forth by those who want to revise history as well as what the Bible clearly declares on this vital subject. May God be glorified as a result of the publishing of this much-needed volume."

-Pastor Gary Small,
Liberty Fundamental Baptist Church,
Lynden, Washington 98264

"As a pastor, I have seen firsthand the devastating effects alcohol so often has on marriages, families and individual lives. One Christian friend told me how alcohol had ruined his life ... and he was not yet thirty! I will never forget him telling me, "I wish I had never been introduced to the stuff." I am very thankful that God removed from my life the consumption of beverage alcohol soon after my conversion in 1980. It has saved me so much grief and allowed my life to shine for Christ more brightly before friends, family and an unbelieving world.

In Ancient Wine and the Bible: The Case for Abstinence, Pastor David Brumbelow does a masterful job explaining the times and customs of Bible days and the scriptural use of the word *wine*. In this book, you will find well-reasoned answers to problematic issues concerning the Bible and alcohol. Is God for moderation or total abstinence? After reading Pastor Brumbelow's book, you will see more soberly than ever the clear answer to that question.

-Jeff Schreve, *Pastor,*
First Baptist Church, Texarkana, Texas

"With the clarity, logic, and thoroughness, an outstanding attorney uses to prepare a brief in a major lawsuit, David Brumbelow approaches the question of drinking alcoholic beverages from a Biblical point of view. This work is outstanding. I recommend it strongly. The upcoming generations need to know the havoc brought on our society and upon individuals by the use of alcohol. If we use it ourselves, we recommend its use to others. A Christian should not exercise his freedom to put himself and others at such a risk."

-Judge H. Paul Pressler,
Justice for the 14th Court of Appeals,
Houston, TX.

Table of Contents

Foreword

The fact that there is "nothing new under the sun" does not imply that there are no surprises. When evangelicals debate the reliability of the Bible, I am never astonished because I know that Satan's first attack must always focus on the reliability of God's Word. But when Southern Baptists spent about thirty minutes at a recent convention debating whether or not Baptists should imbibe, I confess I was bewildered. After years of leadership in the abstinence movement, here was a large evangelical assembly debating the issue.

David Brumbelow was also taken aback by what transpired that day. Having spent years in the pastorate, he had observed the tragedies associated with consumption of intoxicating beverages. Brumbelow determined to research the matter and reassess the teachings of Scripture about alcohol. The volume you hold, *Ancient Wine and the Bible: The Case for Abstinence*, addresses the subject with keen logic, a grasp of history, and thorough exegesis of biblical literature. Acknowledging that the Scriptures do not retain an expressed mandate against drinking alcoholic beverages as "thou shalt not steal," he, nevertheless, demonstrates that the overwhelming witness of the Bible is like a mighty breaking wave on the north shore of Oahu, demanding abstinence based on case histories of the devastation of "strong drink" added to the "wisdom" literature of the Bible in its repeated call for abstinence.

Not only has Brumbelow done his historical and exegetical homework, but also he provides confessional and covenantal evidence from many varied denominations and sources proving the use of "strong drink" unwise. Brumbelow does not shrink from passages that at first blush appear to support the use of alcohol. But he demonstrates that these passages do not constitute an argument for either recreational or dietary use of such spirits. A helpful annotated bibliography is also appended.

No industry, with the possibility of its close partner — pornography, wreaks such devastation on individual lives, on the family, and on the social order. How any thoughtful believer, let alone a pastor responsible for his flock, can support an industry that is the cause of such heartache and sorrow every year is beyond comprehension. A church ought not to call as pastor a man with such limited compassion for his fellows as to drink or endorse those who do. Such is totally inconsistent with the mandate of our Lord for unselfish and even sacrificial love.

David R. Brumbelow's book *Ancient Wine and the Bible* should be consumed by every serious believer, every pastor, and hopefully every seminary student who loves God and wishes to care for the Lord's flock. Legalism is always harmful, but the antinomian loss of sanctification and holiness is by far the most dangerous nemesis of contemporary evangelicals. This book is a purposeful, thoughtful, well-reasoned manifesto for holy living.

Paige Patterson
President
Southwestern Baptist Theological Seminary
Fort Worth, Texas

Dr. Paige Patterson has served as pastor; president of Criswell College, Southeastern Baptist Theological Seminary, and Southwestern Baptist Theological Seminary. A graduate of Hardin-Simmons University and New Orleans Baptist Theological Seminary (Th.M. and Ph.D.), he is the author of a number of books, and was twice elected president of the Southern Baptist Convention.

Introduction

While I already had indication of it, my eyes were opened wide at the 2006 annual Southern Baptist Convention. For the first time in my lifetime, I saw pastors stand on the floor of the convention and openly speak in favor of social drinking. While the resolution against alcohol passed by a wide margin, this indicated that today more people, even some pastors, are in favor of drinking. Gone were the days where even if a pastor hypocritically drank, he would never have stood publicly and advocated such a thing.

Today Christian colleges that once took strong stands against alcohol and other destructive drugs are losing those convictions. More pastors openly condone moderate drinking, are weak on the subject, or ignore it altogether.

Many pastors, youth ministers, and parents who see the danger of beverage alcohol do not have the information and resources to speak up about the issue. This book intends to correct that lack of information and resources. I pray this will be used as a resource to counter those who claim Jesus and the Bible condone the use of a hard drug for no other purpose than selfish enjoyment.

Some claim those who are educated cannot and do not oppose alcohol. That myth is exposed in this book. Many quotes are given by the highly educated, past and present, who oppose alcohol. Reference material will be presented by those from all walks of life. It will be seen that a multitude of the educated, those with little formal education, ancient and modern, preachers, scholars, Christian, non-Christian, have wisely recognized the dangers of intoxicating wine. Brief biographical information of those referenced will often be presented.

I'm weary of instances like the professor who makes fun of the Sunday School teacher who believes Jesus turned water into un-intoxicating wine. Based on the evidence, it just could be the Sunday School teacher is correct and the distinguished professor is wrong.

Throughout this book I will use the term "wine" to refer to that wine which contains alcohol, and that wine which does not contain alcohol. This is the same type word usage as occurs in the Bible, in ancient literature, and even to some extent in our modern day usage. Much evidence for this usage will be presented. You will scarcely find a word in the English Dictionary, but that more than one meaning is given. This is also true of the ancient use of the word wine.

While much of this book has been in progress the last five years, in a sense, this book has taken a lifetime to produce. I had the privilege of growing up under the preaching of a dad who strongly preached against the recreational use of drugs, including alcohol. Joe Brumbelow said he would rather have a rattlesnake in the house than to have alcohol in the home. He preached how the biblical words for wine referred to both alcoholic and nonalcoholic wine. My two brothers and I are all pastors today. We had the privilege of growing up in a home that never included beverage alcohol. As a teenage preacher, I discussed alcohol with Evangelist Dan Vestal. He graciously told me if I ever wrote an article on the subject he would print it in his evangelistic newsletter. He went to be with the Lord in 1980, but I remember still his encouragement.

The *Biblical Evangelist* has printed a couple of my articles on this subject. The *Sword and Trumpet* has reprinted one of them. My thanks to them for their encouragement. Blogs and the internet have often covered this subject. I've been in numerous debates about alcohol on the internet. I've learned much from those debates, even from those who opposed me. (For those who argue minutia, these were not formal, but informal debates.) Peter Lumpkins and Jerry Vines have been faithful in pointing toward little known evidence for this viewpoint.

While my dad went to be with the Lord in 2002, my mother, Mrs. Joe E. (Bonnie) Brumbelow has been a great help, critic, and encouragement to me throughout this study. This process began by mom asking me to publish what became a 17 page article on alcohol to give to her Sunday school class at First Baptist Church, Lake Jackson, TX. My pastor brothers, Mark and Steve, have also been a real encouragement in this endeavor. Thanks to Pastor Gary Small who not only endorsed the book, but offered helpful grammatical suggestions. Thanks to the others who graciously endorsed this book.

My church, Northside Baptist Church of Highlands, TX has been very interested and gracious during my work on this project.

Throughout this book many works are cited. This is not to be mistaken as an endorsement of all works or their underlying philosophies. This book is intended as a strong defense of abstinence from the beverage alcohol. Some falsely assume that abstainers thereby are arrogant and hateful toward any who disagree or whose practices differ. This is not the case. Undoubtedly there are rascals on both sides. As with most debated issues, we have friends on both sides. It is possible and desirable to take a strong stand on what you believe is right, but do so in a loving tactful way (Ephesians 4:15). Drinkers with stereotypical views of Christian abstainers would be surprised at the many pastors, churches, common folks who have lovingly cared for and encouraged those struggling with alcohol.

I have tried to not make this book a personal attack on any one individual. So often the opposing view will be given without directly naming individuals. Rest assured the opposing view has frequently been given in books, speeches, papers, on the internet, and in many a classroom.

This book does not always flow smoothly. Its many quotes and references are intended to provide concrete evidence to the reader. Some of those quotes will be repeated. That evidence is provided to the reader, whether he be formally educated or not. I would much rather be understood by common folk and put evidence at their easy disposal, than for this book to always flow smoothly, be difficult for some to understand, and only impress the elite.

May many be grounded in their biblical convictions through this work. May God use this book to His glory.

Sincerely,
David R. Brumbelow
Highlands, Texas, USA

Dedication

Ancient Wine and the Bible is dedicated to those men and women of the last two centuries and throughout the ages who have opposed, often against great opposition, the recreational use of destructive drugs. You have saved the lives of untold numbers. You have been faithful. You have been wise. To name them would fill volumes. But God knows well every name.

If anyone desires to come after Me, let him deny himself, and take up his cross, and follow Me.

-Jesus Christ; Matthew 16:24

CHAPTER ONE:
Controversy Over Wine in the Bible and in Ancient Times

Many have spoken authoritatively on ancient wine and the Bible. Some have been well educated; some have not. They have often said something like the following:

"The Middle East and Israel is a hot climate. Wine begins to ferment soon after it is pressed out of the grapes. They had no way to prevent this alcoholic fermentation. After all, they had not discovered pasteurization and had no refrigeration or electricity. Therefore, other than for a very brief time, they had no unfermented wine to drink. In addition, their water was unsafe to drink. They had no choice but to drink this intoxicating wine."

Such has been said many times over the years. Assuming the above is true, it presents some serious problems. First, the Israelites, men, women and children, would have been falling down drunk all day every day. After all, they had nothing to drink but alcoholic wine. Second, they would have had to spend an inordinate amount of money on wine. If that is all you have to drink, you would need a lot of it. And it was expensive. Some scholars believe one reason they did not have more trouble than they did with drunkenness in Bible times is because it simply cost too much to do a lot of drinking. "Intoxicants were then [in Bible times] an expensive luxury, beyond the reach of the poorer classes." [1] Third, this assumes the only wine they had was alcoholic, and they had no other option. This is a false assumption. The ancients knew multiple ways to prevent alcoholic fermentation of wine.

While not knowing the details of germs and microorganisms, they did know the importance of clean water. The Bible often speaks of the people drinking water. Ancient people even knew that sometimes it was advantageous to boil water as well as wine. They knew somehow boiling made the liquid clean and safe. However, this writer has never found an instance in the Bible or in ancient literature where the people purified water with wine. The only references found have been modern day references. If ancient people purified water with wine, it seems there would be some reference to it in ancient writings. First Timothy 5:23 commends a little wine (likely unfermented wine) for medicinal purposes; it says nothing about wine purifying water.

Alcoholic Wine

In Bible times they did, of course, have alcoholic wine. Noah got drunk. Lot got drunk. Scripture condemns intoxicating wine. Other ancient examples of clearly intoxicating wine are recorded by the historian Pliny who lived in c. AD 70. He stated, "As for the wines of Pompeii... they are detected as unwholesome because of a headache which lasts till noon on the following day." [2] Furthermore, Athenaeus states that there is a wine "the drinking of which causes insanity among males, but pregnancy in females." [3] Pliny concurs, "One [wine] grown in Arcadia is said to produce ability to bear children in women and madness in men." [4] We still seem to have that kind of wine today! Again Athenaeus observes, "But wine obscures and clouds the mind."[5]

Scripture, as well as ancient literature, however, reveals that the ancients had many kinds of un-intoxicating drinks. They could preserve them in an un-intoxicating condition. The words for wine sometimes specifically referred to intoxicating wine, but significantly, they were also used to refer to un-intoxicating wine. Scripture itself shows this to be true (Genesis 40:11; Proverbs 3:10; Isaiah 16:10; 65:8). Ancient writers such as Aristotle, Hippocrates, Pliny, Columella, and Athenaeus also show this to be true. Many of the most educated, accomplished scholars have recognized that wine in the Bible and in ancient times referred to both intoxicating and un-intoxicating wine. Many of these examples will be provided in the pages that follow.

When grapes are pressed, it is true that at the right temperature wine will immediately begin to ferment into an alcoholic drink - if nothing else is done. If absolutely nothing is done, the wine will probably ferment into something alcoholic that tastes awful and is undrinkable; it will then turn into vinegar or just be completely spoiled. Even drinkable alcoholic wine is no accident; it takes much time and effort. It is also true that good beef from a slaughtered cow will immediately begin to rot and become infested with flies and maggots - if nothing else is done. The people of ancient times knew well how to preserve food and drink; their lives depended on it. They knew that when grapes were harvested, or when an animal was killed to eat, the work had only just begun. Properly processing and preserving the food was more labor intensive than the actual harvest or slaughter. When grapes were pressed, work immediately began to preserve the fruit of their labor. One of the first things usually done was to boil the new wine. Much work had to be done whether this new wine was to be preserved in a fermented or unfermented state.

For some reason modern critics, some of them scholars, seem to think the ancients were clueless when it came to preventing fermentation. The ignorance, however, lies with modern folks, not those of 2,000 years ago. Ancient writers such as Cato, Varro, Pliny, Columella, go into great detail on how to properly preserve meat, grain, fruit, and beverages. They commonly used preservation methods that are almost unknown today. Meat was preserved by drying and with salt. Or they slaughtered the animals as they were needed. Amurca (a byproduct of olive oil) repelled weevils in stored grain. Many types of food were preserved in brine. They practiced what is now called lactic fermentation. Fruit was dried. Fruit was often boiled down to a thick consistency that would keep unfermented and unspoiled for long periods of time. Dried figs and raisins were commonly used. Fruit would be preserved in honey or in thick, boiled down must (unfermented wine). Containers could be sealed airtight with resin, wax, or olive oil. Some fruit and vegetables would keep for long periods of time in cellars, hung from ceilings, or sealed in vessels. The end of the stem of a grape cluster would be dipped in boiling pitch to seal in the moisture and prolong their fresh condition. Preservatives were used such as salt, sulfur, resin, marjoram. Salt and sulfur (sulfur dioxide) are still commonly used today as preservatives.

Meaning of the Word Wine

Many say that everyone today knows the meaning of the word wine. It always refers to alcoholic wine. So when Jesus made wine, it was obviously alcoholic. That statement will be challenged. It is surprising how often wine is used, even today, to refer to nonalcoholic wine. In properly understanding the Bible, however, it is much more important how the word wine was used 2,000 years ago, than how it is used today. Meanings of words vary and change.

As the Bible does, this book will throughout refer to wine in a generic sense. It will refer to the juice of the grape, whether alcoholic or nonalcoholic, as wine. As will be pointed out, the English translations of the Bible also translate the Hebrew and Greek words for wine into the English word wine. They do so even when the wine is clearly nonalcoholic, for example in Proverbs 3:10 and Isaiah 16:10; 65:8. In other words, what we commonly call grape juice is referred to in the Bible as wine.

Sometimes these competing views are called the one-wine and two wine theories. One-wine - that all wine in the Bible is intoxicating wine. Two-wine - that wine in the Bible and ancient times referred to both intoxicating and un-intoxicating wine.

English Words Used for Both Alcoholic and Nonalcoholic Drinks

Just as the Bible used wine in a generic sense, we do the same in our own English language. The English word wine has been used to refer to unfermented wine, although not as much today as years ago. Words we commonly use today can refer to a liquid that is alcoholic or not. Examples include drink, liquor, cider, eggnog, punch.

A preacher says a Christian should not drink. Then he announces a church social next Sunday and says the church will provide the drinks. Is he a hypocrite? Has he taken leave of his senses? No, everyone present understands perfectly what the preacher has said. Drink sometimes refers to an alcoholic beverage; sometimes it refers to a nonalcoholic beverage. The context of his speaking shows what "drink" means.

Liquor often refers to an alcoholic drink, but not always. A dictionary will say that it can refer to any liquid, perhaps a mixed or concentrated liquid. My dad used to drink a little nutritious "pot liquor," the liquid left over after a pot of green beans or other vegetables had been boiled. Mom never liked him referring to it in that way! Chemical companies sometimes refer to a liquor tank because of the chemicals mixed in it. Doctors used to mix a "liquor" for patients that may or may not have contained any alcohol.

Cider refers to the juice of pressed (sometimes called expressed) apples. Cider is used to refer to nonalcoholic apple juice and also alcoholic apple juice. Sometimes it is distinguished as sweet cider and hard cider. Wine can also be sweet or hard. Interestingly, cider comes from an old Hebrew word *shekar*, but more on that later.

Alcohol is sometimes added to eggnog or punch. To some, eggnog or punch always reminds them of an intoxicating drink. To others who would never drink, eggnog and punch refer to perfectly innocent drinks.

Why the Meaning of Wine Matters

Why does it matter whether wine can be either an alcoholic or nonalcoholic drink in the Bible? It matters because we should be interested in properly understanding the meaning of Scripture. It also matters because sometimes the Bible approves wine; sometimes the Bible condemns wine. Does the Bible contradict itself? Of course not. The Bible is the inspired, inerrant Word of the Living God. The answer is simple. The Bible condemns alcoholic, intoxicating wine. The Bible praises wine that is sweet, nutritious, and un-intoxicating.

How can you tell what kind of wine the Bible is speaking of? Usually it is easy. One of the basic rules of understanding the Bible is to study the context. In other words, what has the Bible just been saying, and what does it say right after the mention of wine? For example, Proverbs 20:1; 23:29-35; Ephesians 5:18 are easy to figure out. In these passages it is clear the Bible is speaking of alcoholic wine that will get a person drunk.

Same Words Used Differently in the Bible

Some feel like the Bible only uses a word one way, but that is not the case. The word God (Elohim) is used differently. Sometimes it is used to refer to the one true God, sometimes to false gods, or even to angels. In the Bible the word angel sometimes refers to an angelic servant of God, sometimes to an angelic servant of the devil, sometimes even to a human messenger. The word spirit sometimes refers to the Holy Spirit, sometimes to the human spirit. Would anyone be foolish enough to interpret the words God, angel, spirit in the same way every instance in the Bible? It is also foolish to see wine as the same in every instance in Scripture.

The immediate context of the Bible, as well as the entire context and teaching of the Bible, should determine how we understand and interpret the word wine. For those who honestly consider the evidence, they will no longer see wine mentioned in the Bible as only and always refer-ring to an intoxicating drink.

The Social Drinkers' View of Bible Wine

There is another view of the word wine in the Bible. Some Christian moderate drinkers insist that wine in the Bible and ancient times was always fermented and intoxicating. Therefore, they point to the fact that the Bible sometimes commends wine and that Jesus turned water into wine. Their conclusion is that Christians can enjoy wine and other in-toxicating drinks as long as they don't get drunk. Many add they can even get a little buzzed (slightly drunk), as long as they don't get drunk.

This view of moderate drinkers leaves some unsettling questions. At what point does drinking become too much and bring on drunkenness? Social drinkers have no clear answer to this question. Doesn't the first drink of alcohol affect the mind, begin to dull the senses, and adversely affect good judgment? Pair two athletes of equal ability. Let the first drink a beer, then let them compete; the second athlete will win every time.

If the Bible commends people for moderately drinking wine, which you consider as always a hard drug, this opens the door for moderate use of other drugs like marijuana, cocaine, even heroin. If God approves one hard drug, wine, for recreational purposes, then reasonably He would approve of others. The moderate drinker often says "no," because these drugs are illegal and a Christian is to obey the law of the land. But what about the countries in which these other drugs are legal? Some Chris-tian moderate drinkers admit that if they were legal it would be fine for a Christian to enjoy, moderately mind you, these other hard drugs. I have to at least commend them on their consistency. Many others simply want to avoid such an embarrassing question.

There is something unsettling about Christians saying the Bible ex-cuses the recreational use of any mind-altering drug on the market. Furthermore, the moderationist view is also the view of the saloon keeper, the bartender, the alcohol industry. They are thrilled when Christians take this view. Should a Christian be a supporter of a drug industry that has caused crime, disease, accidents, death, wrecked homes, and untold heartache? The moderationist view often leads to alcoholism. As Adrian Rogers said, "Moderation is not the cure for drunkenness; it is the cause of drunkenness." [6] Most every alcoholic will tell you he began as a moderate drinker. Alcohol often becomes a gateway drug to other destructive drugs. The moderationist view is seriously lacking.

Winemaking in Bible Times

In ancient times knowledge of viniculture was extensive. Innumerable varieties of grape vines were available. Plants and cuttings were transported throughout the Roman world. Vines were carefully planted, grafted, pruned, trenched, and fertilized. Vines were propped up on poles or allowed to grow up into trees. Although many details are not told, vines and wine products are mentioned often in Scripture.

While the general harvest time would have been in the fall, the earliest grape harvest could have been in July and the last harvest in December or sometimes even later. Roman writers referred to early grapes, mid-season grapes, vines with two crops a year, and grapes gathered after a frost. Microclimates such as how much shade or sun vines received, also affected how early or late the crop was harvested. In addition some grapes would be cut from the vines a little early for grapes or wine a little less sweet, or cut at the height of ripeness and sugar content. Grapes were selected for immediate eating right off the vine or as table grapes to be used within days. Some clusters and varieties were good keepers and could be preserved fresh for months at room or cellar temperature. Some of these clusters would be carefully, loosely packed in wicker baskets, packed airtight in various vessels, or hung from the ceiling. These could last throughout the year. They would be eaten fresh or hand pressed into a cup for fresh, unfermented wine (Genesis 40:11).

Some grapes would be preserved in honey, must, or brine. Some grapes were sun dried or smoked into raisins. These raisins could be eaten or re-hydrated back into unfermented wine (or could be then fermented in alcoholic wine). Smoked raisins were a Roman favorite and especially enjoyed by Caesars Augustus and Tiberius.

Grapes to be made into wine would be placed in the winepress. Many winepresses were permanently hewn into rock; others were portable and about the size of a bathtub. Men or women, with hopefully washed feet, would tread the grapes expressing the juice. Usually this wine treading was a community or family event, rather than one having to tread the press alone. The must or new wine would flow from the winepress into a vat. At this point the decision would need to be made as to whether the fresh wine would be preserved in an un-intoxicating state or an alcoholic state. Columella (c. AD 70) said that sweet wine (nonalcoholic) would need to be preserved as soon as it's taken out of the vat; harsh wine (alcoholic) would need to be preserved after several days.

Whether the wine was to be intoxicating or not, it was often immediately boiled. While not knowing the details of microorganisms, ancients did know that boiling wine or water would somehow cleanse it and make it safe. Heating would boil away any alcohol and would kill yeast and any other impurities. A frequent way to preserve the very common sweet, un-intoxicating wine was to boil it down to a thick consistency. The thick, high sugar content would prevent this new wine or *must* from fermenting or decomposing. Aristotle called "must" a kind of wine. Some dictionaries call must "new wine."

When wine is fermented the sugar is consumed by the yeast and converted into alcohol (and gas, carbon dioxide). While the ancients did not have a word for alcohol, they knew some wine would intoxicate; other wine would not. So if wine was sweet, it normally meant it was un-intoxicating wine. Fermented wine had a dry, un-sweet, harsh taste. So the two kinds of wine generally tasted very differently. Aristotle said sweet wine did not inebriate and did not taste like (fermented) wine. But he called this nonalcoholic drink "wine" as he did "must." Fermented wine could later be sweetened, but the common meaning of sweet wine was that which did not inebriate. Harsh or dry wine was the alcoholic kind.

Various preservatives were used for either kind of wine. Nonalcoholic wine was preserved from alcoholic fermentation; alcoholic wine was preserved, often unsuccessfully, from turning to acetic acid or vinegar. Vinegar was called sour wine (Mark 15:36).

Preservatives and additives included salt, sulfur, marjoram, resin, sea water, marble dust, herbs, and potter's clay. Salt or sea water in wine produced lactic fermentation. This actually produced a nonalcoholic fermented wine. Salt and/or whey promote nonalcoholic fermentation. This lactic fermentation appears similar to alcoholic fermentation, but without the alcohol, or with very little alcohol. Sulfur is still used today as a food preservative. Sulfur could keep unfermented wine from fermenting or keep fermented wine from souring to vinegar.

The length of time the wine was left on the lees (the sediment, grape skins in the vat) affected the taste, color, and clarity of the final product. Flavors of both alcoholic and nonalcoholic wine could be affected and enhanced by age. Most wine was consumed within a year of its production. As today, much praise of certain wines was exaggerated.

It was very common in Classical times to add three to five or more parts water to wine. For intoxicating wine, this would greatly reduce its inebriating effects. For sweet wine it was to reconstitute it to a drink that was not too strong or syrupy. This reconstitution was much like today making a drink out of snow cone syrup or adding water to lemon juice for lemonade. Aristotle referred to wine so thick from smoke it had to be scraped from the wineskins.

Wine was often used as a medication or as part of a medical mixture. This not only included fermented wine, but also unfermented wine and the other grape products listed above. Grape products were somewhat of an ancient multi-vitamin. Wine was used as medication both internally and externally.

Wineskins were made from processed animal skins. Some wineskins were made from a whole ox hide or goat hide, thereby making a large, light, collapsible container. An animal could be skinned (or cased) in one piece, where the only openings were the feet and neck. The feet openings would be sewed back watertight. When processed these wineskins would travel more easily on the back of a camel or donkey. New wine was not put in an old wineskin, but not so it would not burst. Fermenting wine would burst a new or old wineskin. Rather new wine or must would not be placed in an old wineskin to keep it from contamination and ferment. Cleanliness was crucial.

Intoxicating wine was not the major product of Israeli vineyards. Sweet un-intoxicating wine in various forms, fresh grapes, raisins, raisin cakes, preserves, were very common and used as food, drink, and flavoring.

Un-intoxicating boiled-down wine is still used and is called by different names in different countries. Names include pekmez, sapa, saba, dibs, vin cotto (cooked wine), grape molasses, must, etc. It is used for preserves, to flavor and sweeten foods, or water can be added for a non-alcoholic wine or drink. While cane sugar (not available in Bible times) sweetens without adding flavor, grape molasses users boast their product both sweetens and flavors. They also point out the health benefits of this boiled-down wine.

Wine is referred to in Scripture as grapes on the vine, fresh expressed grape juice, un-intoxicating wine, intoxicating wine, and vinegar. This same wide general use of the word wine was used not only in the biblical languages of Hebrew and Greek, but also in the neighboring languages of Egyptian, Hittite, Sumerian, Akkadian and Latin. Wine was used in this generic way in Greek and Roman literature. Some ancient writers would at times indicate wine was only the intoxicating kind, often later contradicting themselves by calling must, sweet wine, or boiled wine, "wine."

Aristotle, Hippocrates, Polybius, Athenaeus and others pointed out the different properties in different kinds of wine. They wondered why some inebriated and others did not, why some tasted like wine and some wine did not, and why some tasted sweet and some tasted harsh. They noted that some wine got thicker when it was boiled and some did not, some wine was flammable and some was not, some wine easily froze but other wine did not. The even wondered why some wine led to addiction and some did not. It is much easier for us to understand today. The Bible did not have a specific word for alcohol, but it described it in Proverbs 20:1 and 23:29-35. The Bible condemns this kind of wine, yet commends the un-intoxicating wine, or grape juice, that was used as a food rather than a drug.

Today we have a word for alcohol, know microbiology, and know why some wine is intoxicating and some is not, why some is dangerous and some is not. In ancient times they identified it by descriptions of the effects of intoxicating wine.

Kinds of Wine Vary

There are numerous kinds of wine (Nehemiah 5:18), including numerous kinds of alcoholic wine and nonalcoholic wine. Ancient literature refers to all these kinds of wine. Pliny records, "Who can doubt, however, that some kinds of wine are more agreeable than others, or who does not know that one of two wines from the same vat can be superior to the other, surpassing its relation either owing to its cask or from some accidental circumstance?" [7] This would have been true of either alcoholic or nonalcoholic wine. Pliny refers to this again in another text, "We have already described the various kinds of wine, the numerous differences which exist between them, and most of the properties which each kind possesses." [8] One final example from antiquity comes from Vitruvius Pollio a Roman writer, who says, "But we find in the island of Lesbos the protropum wine, in Maeonia, the catacecaumenites, in Lydia, the

Tmolian, in Sicily, the Mamertine, in Campania, the Falernian, between Terracina and Fondi, the Caecuban, and wines of countless varieties and qualities produced in many other places." [9]

Alcoholic wine was far from the only, or major, product of the vineyard. A modern example of this shows how grapes in Turkey are used. "Of the grapes produced in Turkey, 24% of them are eaten fresh, 35% are dried into raisins, 3% are used for wine, 20% are used to make molasses, and 17% to produce *pestil*, or fruit leather." [10] The vineyards in Turkey today are the fourth largest in the world. These percentages may have been similar in Bible times. The 20% made into molasses could have been drunk as unfermented wine by simply mixing it with water.

Difference in the Taste of Wines

Alcoholic fermentation turns the sugar content of wine into alcohol and carbon dioxide, leaving the wine tasting dry or un-sweet. New unfermented wine is very sweet to the taste because it has not lost its sugar content to fermentation. Wine preserved through salt and / or whey, and lactic fermentation would also be sweet.

While wine connoisseurs may protest, let's make the difference very simple. As a general rule, in ancient times nonalcoholic wine was sweet; alcoholic wine was dry, harsh, or un-sweet. A rough example would be the difference between drinking sweet tea verses drinking un-sweet tea. Of course, in the case of tea, neither would be alcoholic; this is only referring to the general taste. So the first taste could immediately tell them the nature of the wine.

Some will argue there was and is alcoholic sweet wine. That is true, and there are exceptions. Wine can be fermented, then a sweetener added. As far as that goes, any drink can be made alcoholic. Diet Dr. Pepper® is a popular, un-intoxicating soft drink in Texas. But anyone can add alcohol to it and make it an intoxicating beverage. It is still correct to assume Diet Dr. Pepper® is a soft drink, rather than a hard drink.

Wines of today, whether sweet and/or intoxicating, cannot properly be assumed the same as ancient wine. Let the ancient writers define the various wines themselves. They are greater authorities on wine in Bible times than drinkers of today. Ancient wine was much different than today. The normal, natural meaning of ancient sweet wine or new wine is that it was the unfermented, un-intoxicating kind of wine. Aristotle, Columella, and Hippocrates confirm this view.

Sweet, un-intoxicating wines can age and gain flavors, just as alcoholic wine can do. Those who preserve nonalcoholic wine, whether by lactic fermentation or filtration, speak of how the flavors are enhanced over time. Aristotle said that sweet wine does not taste like (alcoholic) wine.

Draper Valley Vineyard is a maker of 100% grape juice (unfermented wine) out of wine grapes (Pinot Noir, Chardonnay, Riesling, Cabernet Sauvignon, Muscat, Gewurtztraminer). The owner has said the only ones that object to the taste of his product are those who expected it to taste like alcoholic wine. In their advertisement, they even say their grape juice does not taste like wine. But the tastes are outstanding.

In contrast, most "nonalcoholic wine" is dealcoholized wine. It is made from fermented, alcoholic wine that is then filtered to remove the alcohol. This nonalcoholic wine would taste like alcoholic wine. Wine connoisseurs often say the dealcoholized wines taste terrible. However, Ariel, a dealcoholized wine, boasts their wine was secretly entered into an alcoholic wine contest and won the gold medal. The *Oxford Companion to Wine* states alcohol can be removed without necessarily "harmful effects on flavour and quality."

Most every drinker will tell you that at first they did not like the taste of alcoholic wine. It is an acquired taste. Interestingly, it also seems that way with other drugs like beer and cigarettes. Could our bodies be telling us something? On the other hand, it seems everyone loves the taste of fresh, unfermented wine or grape juice. You can enjoy it the first time you drink it.

Sweet Wine

When ancient literature refers to sweet wine, it is normally a reference to nonalcoholic wine, or what most today would call grape juice. It is not to be assumed the same sweet wine or dessert wine of today. Ancient writers confirm sweet wine was un-intoxicating. Aristotle said sweet wine would not inebriate and did not taste like alcoholic wine.[11] Hippocrates said sweet wine affects the head less, attacks the brain less.[12] Plutarch held, "Wine should not be heady till it hath lost its sweetness."[13] Pliny maintained "Sweet wine...is less inebriating."[14] Athenaeus, "Now sweet wines do not make the head heavy."[15] In a letter Plato wrote he was sending the children sweet wine and honey. [16] As today, they knew alcoholic wine was improper for children. This is reminiscent of the Scripture telling of infants crying for wine. Parents don't give their children alcoholic wine; they do give their children and infants fruit juice.

Alcoholic wine could be made sweet by adding concentrated sweet wine. Sometimes you will come across an ancient reference to sweet wine that was intoxicating. Any beverage can be made alcoholic. But alcoholic sweet wine was the exception rather than the rule.

Just pressed wine starts out extremely sweet. Just as today, ancient people craved sweet things. Unlike today, sweets were very limited. Fruit and honey were their only sources of sugar. Aristotle and Theophrastus mention the craving people have for sweet flavors. The alcoholic fermentation process took away the sugar content; so sweet wine was now bitter or at least lacking the previous sweetness.

Many craved sweet wine because of its taste, and because it was non-intoxicating and non-addictive. Sweet, nonalcoholic wine had a taste distinctive from alcoholic wine. So when someone says sweet wine did not taste like wine, that is a further indication of un-intoxicating wine. Though they had no word for alcohol, it was easy to distinguish alcoholic wine. The first sip revealed whether it was dry alcoholic wine or sweet nonalcoholic wine. When wine is referred to in ancient literature as sweet, it should be assumed it was nonalcoholic wine unless the context indicates otherwise.

Mixing Wine with Water

Ancient wine was commonly mixed with three to five parts water, sometimes even more, up to 20 parts water. Dr. Nelson Price explains, "Writers normally referred to wine mixed with water as 'wine.'" [17] This is well documented; for example, "Homer has stated that the Maronean wine was mixed with water in the proportion of twenty measures of water to one of wine. The wine that is still produced in the same district retains all its former strength, and a degree of vigour that is quite insuperable. Mucianus, who thrice held the consulship, and one of our most recent authors, when in that part of the world was witness himself to the fact, that with one sextarius of this Wine it was the custom to mix no less than eighty sextarii of water: he states, also, that this wine is black, has a strong bouquet, and is all the richer for being old." [18] Some have said this was fantasy in Homer's tale. But years later Pliny says that type wine was still produced. It must have been a strong, boiled-down or inspissated nonalcoholic wine. The mixture of water with wine was common as indicated, "At a symposium in Greek style there was one universal feature, the krater, Latin cratera, in which the communal mixture of wine and water was made." [19]

This is not done today which is another example of the fallacy of comparing modern wine to ancient wine. Here are six reasons why wine was mixed with water in the ancient world.

1. Thick wine was mixed with water to reconstitute it. New wine was commonly boiled down to a thick consistency. This thick wine would be preserved without refrigeration and would not ferment. But it was too strong, rich, and syrupy to drink undiluted. While not alcoholic, drinking this wine too strong could make you sick. Similarly, Scripture says not to eat too much

honey (Proverbs 25:16,27). A substance does not have to be alcoholic to make you sick. Some wine was boiled down to half its consistency, some a third, some even more. Some wine was so thick it was a solid; this would require a large amount of water to properly mix it for a beverage.

2. Even some fresh expressed wine was very sweet and rich. Sometimes this new wine would be mixed with water to achieve the right taste. This would also allow the new wine to go further. For example, the juice of a lemon is too strong to drink. So lemonade is made by mixing a little lemon juice with water and sugar. Some prefer adding water to the premium grape juices of today.

3. Raisins were soaked or boiled in water to reconstitute them and pressed into wine.

4. Alcoholic wine was mixed with water to make it less intoxicating and not as hard on the stomach. It could also be argued that this allowed them to drink more. Those who drank neat (undiluted) wine were often considered barbarians. Romans generally viewed neat wine as poison. A number of Alexander the Great's soldiers are said to have died from a neat wine drinking contest. "It was widely accepted that unrestricted drinking of neat [unadulterated] wine would cause madness and death." [20]

5. Honey wine is mentioned in ancient literature. It was made simply by mixing honey and water. This is another example of a substance being too rich to drink, so they would dilute it with water. Some scholars contend that thick, boiled wine was sometimes referred to as honey.

6. Concentrated wine was easier to carry when traveling. When the traveler stopped to rest, he could mix his concentrated wine with water from a local source. This would lighten the load and make travel much easier.

What about Isaiah 1:22

"Your silver has become dross, your wine mixed with water."

-Isaiah 1:22

Mixing wine with water to drink was widely and commonly practiced during Bible times. It was improper, however, to sell wine mixed with water while claiming it was the unadulterated unmixed product. It would be adding a filler, while claiming it was a pure product. The product was usually not mixed with water until ready to drink (other than seawater, explained later). This Scripture is saying their wine has become adulterated, is no longer pure. Similarly, their silver has lost its value and become dross.

Wine Purifying Water

People of Bible times did not have to drink wine because the water was unsafe. The idea of purifying water with wine was unknown in ancient times. Drinking water is mentioned numerous times in the Bible and ancient literature. Pliny said water was the most healthy of beverages to drink. A few examples of drinking water in the Bible: Genesis 21:14, 19; 24:14-19; 26:18; Exodus 7:21-24; Deuteronomy 2:6; Judges 7:6; 1 Samuel 30:11; 2 Samuel 23:15; 1 Kings 17:4, 10; 2 Kings 3:17; 2 Kings 6:22; 2 Kings 18:31; Job 22:7; Psalm 110:7; Proverbs 5:15; Proverbs 25:21; Isaiah 43:20; Jeremiah 2:18; Lamentations 5:4; Daniel 1:12; Amos 4:8; Jonah 3:7; Matthew 10:42; Mark 9:41; John 4:13.

The people back then knew how to find clean water. They knew the cleanest water usually came from springs. They dug and cared for water wells. They also occasionally boiled water, just as they often boiled wine. They did not know the details, but they knew that boiling a liquid somehow cleansed it and made it safe. Romans often mixed sea water with wine. They were advised to get the sea water as far from land as possible. They knew this sea water would be cleaner and more salty.

Their water may not have been quite as clean and disinfected as our modern day processed water, but it was basically clean and safe. They likely built up a resistance to some water-borne microorganisms, just as some Americans go to another country and get sick drinking the water the locals drink on a regular basis.

No Fortified Wine in Bible Times

Occasionally an authority will speak of ancient people fortifying, or strengthening the alcohol content of wine. Fortified wine, however, was not used and not a possibility for the ancient world. The process of distilling alcohol was not known and used until centuries after Bible times. So the ancients could not add alcohol to their wine. Andrew Dalby explains, "No technique practiced in ancient times made it possible to increase the alcohol level of wine above what is attained in fermentation." [21]

Sometimes drugs were mixed with wine thereby making it more intoxicating. But the common alcoholic wine was not as strong as much wine today, and certainly not akin to fortified wines and strong alcoholic drinks of our modern world. The highest possible alcohol content of ancient wine was around 15 or 16 percent. Most fermented wine would have been much lower. Also, the alcohol content was reduced further by the almost universal practice of mixing wine with water. Usual mixtures were three to five parts water to one part wine. While it could be expensive, it was certainly possible to get drunk in ancient times. There are many instances where they did so. It is a huge error, however, to assume all wine in ancient times was intoxicating.

Some make an issue of the Hebrew word shekar (Greek sikera) which is sometimes translated "strong drink." However, this refers to a similar drink to wine but made out of fruit other than grapes. Like wine, shekar may or may not be an intoxicating drink (see chapter 8). We get our words sugar and cider from shekar. But shekar does not refer to a strong, fortified drink.

Ancient Wine and Personal Tastes

When it comes to ancient wine, it should be pointed out that tastes change. What one country treasures in cuisine, another country finds distasteful. Tastes also change over the years. What tasted good to one generation is rejected by another generation. In addition, tastes are acquired. Some food and drink may taste objectionable at first, yet if you grew up on it, or if forced to eat or drink it, you begin to like it. Resinated wine, for example, is very distasteful to many, but others prefer it. So the wines of ancient times may or may not be appealing to our modern generation. That is not the issue. The issue is what kinds of wine they had in Bible times. The issue is not what kinds of wines are preferred today, but what kinds of wines were preferred by the people of Bible times. The wines ancients preferred were often un-intoxicating, sweet, or filtered, or heavily diluted with water.

A Pro-Drinking Argument

A publisher of a pro-drinking book uses the Bible to defend drinking. Their advertisement asserts, "Some people claim that in Bible days, the wine was unfermented. If that were true, how did Noah get drunk from the grapes of his vineyard? (Genesis 9:20-21) There is no such thing as unfermented wine. The yeast exists on each grape berry (the dust you see on each grape), and fermentation starts the moment the berry skin is broken. Unfermented grape juice can exist easily if pasteurized. Even refrigerated juice will ferment over time, as will most cold things - just look in the back of your refrigerator at the stuff you have not thrown out for months."

This statement is false for multiple reasons.

1. It implies abstainers believe all wine in the Bible was unfermented. This is not true; they only say that not all wine in the Bible was alcoholic. Just as today, the ancients had both alcoholic and nonalcoholic drinks. To say all wine was alcoholic

because of Noah would be like a future historian pointing to someone drunk today and thereby concluding all drinks in our day are obviously intoxicating. The truth, however, is that some are intoxicating and some are not.

2. There is such a thing as unfermented wine and numerous examples are given in this book.

3. To argue that because yeast is on the skin of the grape, that made it impossible to prevent fermentation, is like arguing that when they killed a cow or sheep, it was impossible to keep the meat from rotting. Ancient people were much more resourceful and intelligent than the unnamed person above gives them credit.

4. New wine will decompose and ferment if nothing is done to it. Further, if nothing is done, the fermented alcoholic wine will next decompose and ferment into sour wine, acetic acid, vinegar. The ancients could make intoxicating wine, and they could make sour wine. But they also knew how to preserve un-intoxicating wine, often called must, new wine, or sweet wine. It takes some effort, however, to make and preserve any of these three kinds of wine. The above argument is impressive only to the uninformed. The unnamed author is either ignorant of these ancient practices or is being deceptive to win his argument.

Word Usages, Meanings and Historical Linguistics

The same word can have varied meanings. Consider the following discussion by ancient food authority Andrew Dalby on the terebinth tree and fruit: "Did the young Persians eat 'turpentine wood'? It appears so from B. Perrin's translation of Plutarch's *Life of Artoxerxes* section 3.2 (Plutarch, Parallel Lives tr. B. Perrin, vol. II). You might say, 'We don't eat turpentine wood, therefore the Persians didn't.' This is an invalid argument: if you used the same argument to decide whether Greeks ate cicadas your conclusion would be wrong. But you might, after investigation, say, 'No humans anywhere can be shown to eat turpentine wood, therefore the Persians probably didn't.' This is a valid use of comparative evidence. It is not conclusive but it should impel you to look hard at the source of the dubious information. This is where you gain if you can be independent of translators. Does the error (if that is what it is) lie with Plutarch or with his translator? You won't know until you look at the Greek words. You will then find that the word terminthos meant not exclusively the wood, but also the fruit or nut of the tree (which is more often called terebinth than turpentine, incidentally) and that people in the near east do indeed gather and eat its tiny oil-rich fruits; if you have traveled in the region you may well have seen them on sale at the street stalls. Common sense assists you to the conclusion that Perrin has indeed made an error and that the young Persians ate terebinth fruits, not terebinth wood."[22] In the same way, we should be careful in how we translate and interpret words like wine.

We should also consider that experts can sometimes be wrong. Dr. Robert P. Teachout correctly analyzes scholars and dictionaries, "Not only have contemporary Bible scholars failed to consult the older reference works which would refute their superficial conclusion; they have also neglected to do adequate research in the primary sources themselves. Instead of making a person search through the classical Greek and Latin writings to see (for example) how oinos and vinum were used by the ancient authors they have simply assumed that the modern lexicons [wordbooks, dictionaries] are correct. It is initially important to recognize that lexicons are, ideally based on such primary research. However, lexicons are not inerrant. If the research on which they are based is not thorough, or if it is biased, then important information is omitted. Thus, an adequate investigation of the meaning of any word (especially if the meaning is disputed) must include actually studying it in the context in which it occurred in the original literature rather than simply assuming that the modern lexicons have done a thorough job of presenting the evidence. The difficulty of doing such primary research should make those who take an opposing position more careful about being dogmatic concerning matters, which they themselves are unwilling or unable to thoroughly investigate." [23]

In the following chapters, this book will consult the scholars, all the while going back to the original sources to specifically see how ancients used the words for wine.

CHAPTER TWO:
Ancient Methods of Preserving Non-Alcoholic Wine

A Christian speaks against alcohol and explains how the biblical words for wine were used to refer to non-alcoholic, as well as alcoholic wine. A scholar replies, "But it was impossible to keep wine from fermenting in the ancient world. No one could do this until Louis Pasteur and Welch's in the late 1800's." He adds for good measure, "The Passover wine had to be fermented because it was in the spring, long after the fall grape harvest." [24] A long time professor argues that, "Only in the last 100 years with a sterile environment and chemical additives has fermentation been postponed. The ancient world could not stop the natural process of fermentation." [25] This false assumption is oft repeated from pulpits to entire congregations. Another example, "Passover is a full SIX MONTHS after the grape harvest. There is literally no way to have 'fresh grape juice' at a Passover meal without pasteurization, which wasn't invented until the 19th century." [26]

These quotes are typical of many learned professors and seminary graduates, those who teach preachers and write commentaries. Those who use this argument think they are rightly interpreting Scripture. Their arguments seem unanswerable. But are they? No! Sadly, those who make these arguments are taking their own ignorance and projecting it onto the Bible and the ancient world. To argue that the ancients could not preserve un-intoxicating wine is wrong factually, scientifically, historically. Actually, fermented wine was more difficult to make and preserve than unfermented wine. The problem with fermented wine was its tendency to further ferment into acetic acid or vinegar (which they also called "wine"). "It is estimated that early Romans lost as much as 10% of their wine to the bacteria that causes wine to turn to vinegar." [27]

They also had problems with alcoholic wine picking up bad flavors and smells.

The Bible says very little about how to preserve food and drink. However, these practices were well known and widely practiced. Similar to wine, the Bible often speaks of oil. Yet the Bible says little about the care and grafting of olive trees, gathering, producing and preserving olive oil or its useful byproduct amurca. Nevertheless the biblical people obviously possessed this knowledge. The Bible speaks of milk, cheese, butter, honey, beef, mutton, cabrito, grain, fruit trees; but it says little to nothing about how they were produced and preserved. Scripture says little to nothing about preserving wine, whether fermented or unfermented. It speaks often of agriculture, but is not a how-to handbook on agricultural practices.

The people of Bible times were accomplished farmers and gardeners. They lived in an agrarian society. They were well prepared to grow, harvest, process, and preserve crops of fruit, wine, grain, oil, as well as meat products.

There were numerous varieties of grapes. Trees, vines, and crops were propagated by cuttings, suckers, seed, and by grafting. Varieties of grapes were known for their characteristics of growth, time and length of harvest, whether they kept or conserved well, whether they produced one or two crops a season, type and quality of grapes. Grape vines and cuttings were transported throughout the Roman world. Israel was at the crossroads of the world. Agricultural products, including grape cuttings, would have flowed through Israel from the Mediterranean Sea, Europe, Asia, Mesopotamia, Egypt, North and Central Africa. A new variety of grape would have been sought after and been big news in that day. "By classical times [7th century BC - 5th century AD] vine cultivation had reached a high level of skill: details are given in Latin agricultural writers." [28]

Even down to modern times, Israel is known for their accomplished agriculture. I remember years ago hearing a Southern Baptist missionary telling of agricultural missions. He said, however, there are two countries where they are so accomplished in agriculture that we usually can't help them; they help us. Those two countries were Israel and Japan. Those who are sorely lacking in this historical knowledge have a tendency to project their lack of agricultural knowledge onto the people of Bible times. That makes for a poor understanding of the Bible and is dismissive of the extensive skills of the people of the ancient world.

Unfermented wine could easily be preserved without electricity, refrigeration, or pasteurization. The ancients had at least nine different ways to insure against fermentation and to preserve unfermented wine. Let's explore them.

1. Boiling Down Fresh Expressed Wine to a Concentrated Form

It was very common in ancient times to boil fresh expressed wine or grape juice down to about one third or one fifth its consistency. Sometimes it was boiled down even more. This thick, strong wine or syrup would keep well without fermentation. It is so thick and sweet that microorganisms cannot attack it. When ready to drink, it would just be mixed with water. This was also done with cider and other fruit. Sometimes this thick wine would be used as sweeteners or used like fruit preserves. But it was also used as a drink. This practice not only preserved wine in an unfermented state; it also made it easier to be used by the traveler. Instead of carrying the entire weight, he would carry the concentrated wine and then add local water when ready to drink.

Boiled down, concentrated wine was used in ancient times and is still used today. It has gone by various names including vin cotto (cooked wine), mosto cotto, pekmez, saba, sapa, dibs, defrutum, must, grape molasses, reduced wine, wine syrup, and simply "wine." Vin cotto (or vino cotto, vincotto) is one example of this type of preserved wine being called cooked "wine" even though it is not alcoholic. Balsamic Vinegar is made from cooked or reduced wine. This boiling process concentrates the sugar. Even those wishing to make alcoholic wine have said, "If the sugar concentration level of the must becomes too high at any given point - either at the beginning or during the fermentation - it starts to have an inhibiting effect on the yeast's ability to produce alcohol. In other words, the higher sugar concentration starts to act as a preservative effecting the fermentation in a negative way." [29] They realize that if the sugar content gets too high, it inhibits fermentation. The ancients also knew this and used this process to preserve unfermented wine or grape juice.

Ancient evidence of boiling down wine is extensive. As previously indicated, the word *must* speaks of new, sweet wine or grape juice. The best must comes from certain types of grapes. Andrew Dalby mentions, "*Helvola* or *varia* or *variana* was variable in colour. The darker grapes gave the largest quantity of must...Cato DA 6; Columella DA 3.2.23; Pliny NH 14.29" [30] The ancient writer Columella writes, "The more the must is boiled down - provided it be not burnt - the better and the thicker it becomes." [31] He continues, "The bailiff should have in readiness logs which he may use for boiling down the must to a third or half its original volume." [32] Varro, c. 36 BC, often mentioned boiled must, and preserving fruit in boiled must. [33] Boiled must is just another form of preserving unfermented wine. Cato (c. 270 BC) refers to boiled wine, "Anician frost-pears (these are excellent when preserved in boiled-wine)," and Orcite olives were preserved by "pack[ing] them in boiled must without salt." [34]

Aristotle is quite an authority. He said the wine of Arcadia was "so dried up in its skins by the smoke that you scrape it to drink ... There is a kind of wine, for instance, which both solidifies and thickens by boiling - I mean, must." [35] This possibly solid wine would be mixed with a comparatively large amount of water to drink.

When boiling down must, the scum that rises to the top periodically needs to be skimmed off.

Speaking of the work of a farmer's wife, the Roman poet Virgil (70-19 BC) records, "With Vulcan's aid boils the sweet must-juice down, and skims with leaves the quivering cauldron's wave." [36] Later he writes, "And rose-leaves dried, or must to thickness boiled by a fierce fire, or juice of raisin-grapes." [37] Notice that Columella and Virgil are writing during New Testament times. Aristotle lived at the close of Old Testament times.

Returning to the writings of Pliny we see, "As to siræum, by some known as "hepsema," and which in our language is called 'sapa,' it is a product of art and not of Nature, being prepared from must boiled down to one-third: when must is boiled down to one-half only, we give it the name of 'defrutum.'" [38] These designations were not strictly kept. Often they simply called it wine, must, etc.

More recently writers and speakers demonstrate their knowledge of the ancients' practice of preservation. Jim Richards states, "Several techniques were practiced to prevent or delay the fermentation process. Storage in a cool place extended the life of grape juice. This could have been done in caves and wells. Boiling prevented the fermentation of grape juice. Wine was diluted for consumption."[39] Andrew Dalby speaks of *caroenum,* or grape syrup indicating, "This same region [Lydia] was later the source of an expensive grape syrup, caroenum Maeonium."[40] A Latin Dictionary refers to this Carenun (or Caroenum) as "a sweet wine boiled down one third." [41] Note this dictionary calls this boiled syrup "wine." Tischendorf wrote of a visit to Coptic monasteries in Egypt in

1845, "Instead of wine they use a thick juice of the grape, which I at first mistook for oil." [42] Dr. Price states, "Ancients had several ways of preserving unfermented wine. One way was to reduce the grape juice to the constituency of a thick syrup or even jelly known in Hebrew as debhash and in Arabic dibs. This preserved form could be used over a long period of time. By adding water the concentrate turned the water to unfermented wine. Sometimes a cake was made of dried grapes which later had water added to produce unfermented wine." [43]

Patrick E. McGovern is a pro-drinking secular authority on ancient and modern wine. He said, "Concentrating grape juice down by heating is still used to make the popular *shireh* of modern Iran and was known to the ancient peoples of Mesopotamia as well as the Greeks and Romans. It enables fruit to be preserved, and, diluted with water, it produces a refreshing, nonalcoholic beverage." [44] This is a very notable statement by one of the foremost authorities on wine.

Richard Teachout has served as a missionary in Central African Republic and France. "When I was invited into a French home in Africa, I was often served a delicious drink where a quarter of an inch of concentrated grape syrup was poured into a glass which was then filled with water." [45]

A modern equivalent of this practice can be illustrated in the preservation of cider. Cider, of course, is apple juice. The word cider, like wine, can refer to un-intoxicating sweet cider or intoxicating hard cider. Annie Proulx's 1980 book gives directions for preparing Cider Concentrate or Boiled Cider. You prepare non-alcoholic sweet cider basically the same way ancients prepared unfermented wine, boiling five gallons of fresh cider down to a one gallon concentrate. Later, when you are ready to drink it, simply add four or five parts water to one part concentrate. Further there is a recipe for Boiled Cider Jelly. Just boil the sweet, fresh cider down to one seventh its original consistency. These directions could have as easily been written in 80 BC, as AD 1980. [46]

Such a concentrated juice keeps very well without fermenting. It's the same principle as honey being so thick it does not need a preservative to keep well. Ancient people used these very same principles with wine, and modern cooks are still using such unfermented wine today.

Even the New York Times reveals, "Centuries ago, when refined cane sugar was a luxury for aristocrats only, the common folk made sweeteners by concentrating the juices of grapes and other fruit into dark, viscous syrups with a touch of acidity...Vin cotto means 'cooked wine' though made from unfermented grape juice, or must, and contains no alcohol...Ms. Kasper notes in 'The Italian Country Table' (Scribner) that farmers in Italy still make vin cotto or saba, though sugar is now cheap. One farm woman told her: 'Sugar sweetens. Vin cotto flavors.'" [47] Notice that vin cotto means cooked wine, even though it has never contained alcohol.

We can read about modern chefs reverting back to this ancient practice: "Sapa or vino cotto is the Italian name for grape molasses, and the dark syrup is used in similar ways in Italy. Modern chefs from all over the world have started to experiment with this precious old-fashioned ingredient. I hope that more wineries will start to supply the market with more petimezi (or vino cotto) that has been proven to be not just delicious, but also a healthy sweetener...[also] "Petimezi (grape molasses), is an old-time classic ingredient of the Greek pantry. It is produced by cooking down the grape must for hours, until it becomes dark and syrupy. Petimezi keeps almost forever, and it was one of the ancient sweeteners, together with honey. Its flavor is not just sweet, but much more complex, with slight bitter undertones. The lengthy process makes petimezi expensive, and hard to find, as the small quantities produced every fall, disappear fast." [48]

Some doctors speak of the value of this boiled down unfermented wine saying that grape molasses has more nutritional value than honey. They also point out that it has a big importance for pregnant women, babies, and children, as it is rich in calcium and potassium. It's a great source of energy. They say that 200gr molasses is equal to 1550gr milk or 350gr meat in terms of calories they contain. [49] Likewise, "Grape Molasses is a good source of energy and carbohydrates because of its high sugar content. It is ideal for daily doses of calcium, iron, potassium and magnesium. Plus, it contains lots of minerals. You can purchase Tahini (Tahin) and Grape Molasses (Pekmez) from Turkish or Middle Eastern Grocery Stores or online." [50] I've purchased pekmez or grape molasses over the internet. It was made in Lebanon and is about the consistency of honey. I've used it mixed with water for unfermented wine or grape juice, and mixed it with milk (Song of Solomon 5:1). Mixed with milk it tastes a little like a milkshake. It makes an interesting nutritious drink they used in Bible times. Try it yourself.

One online article explains that the natural sugars in saba act as a preservative to prevent it from going bad. An unopened bottle can keep for years. It is now recommended that an opened bottle should be refrigerated, however, some keep opened bottles un-refrigerated and have it last for months. I have a jar of grape molasses that, after it has been opened and partially used, has kept well for months at room temperature without fermentation or spoilage.

Pekmez is one of a number of names referring to boiled down, concentrated wine. It was developed by the Turks in order to preserve grape juice into a long-lasting form and helped meet the need for sugar and sweetener. "Pekmez is a sweet thick liquid made by boiling and concentrating fruit juice. It is most often made from grapes, but there are local forms of pekmez made from other materials including mulberries, plums, apples, pears, sugar beets, watermelon, sorghum and pomegranates…In Turkey, pekmez and fruit leather are made everywhere there are vineyards. Pekmez is known by different names in different regions; some of the most notable names are *Zile pekmezü*n Zile; *ağda* in Gaziantep; *çalma* in Kırşehir; *bulama* in Balıkesir; and *masara* in Maraş. Pekmez is made from late September to early October when grapes are ripening, and this time is commonly known as 'pekmez time.' The must obtained by pressing the grapes is known as *şira*. The same terms applied to must that has been subjected to a certain amount of fermentation. Pekmez is eaten fresh as well being used to make other products such as *köfter*, *cevizli sucuk* (walnuts strung on a thread and dipped into thickened pekmez) and *pestil* (fruit leather). It is also used as a sweetener in place of sugar in dishes such as fruit compotes. We also know that pekmez was mixed with water and drunk as a beverage." [51] Notice it is sometimes mixed with water to make a nonalcoholic beverage or wine. Also notice sira, like the word wine, can refer to unfermented or fermented wine. Turkey is the fourth largest nation in terms of acreage in vineyards. Its largest uses of grapes are for fresh eating, drying (raisins), and pekmez. In ancient times the area of Turkey was called Anatolia.

2. Long Grape Harvest.

The major grape harvest occurred in the fall of the year. But the entire grape harvest could easily last six months or more. This was done by planting different varieties of grapes that ripened at different times. This was also accomplished by planting grapes in different micro-climates. Micro-climates would make the same grapes ripen a little sooner or a little later. Micro-climates are determined by the amount of sun or shade, whether that exposure is to the south or west, whether the grapes are planted at the bottom, side, or top of the hill.

Ancient writers testified of the vast number of varieties of grapes; some said they were innumerable. The Israelites were deeply involved in agriculture and knew the varieties of grapes and details about viniculture. Some vines bear an early harvest, some mid-season, some late. The first grapes can be picked as early as July, the last in December, sometimes even later. Some grapes were best after a frost. Some vines ripen all their grapes at once; others over a long period of time. Some grapes bore two crops a year.

The result of this lengthy season meant that fresh grapes right off the vine were available for half the year or more. These fresh grapes could easily be made into new wine by simply squeezing them by hand into a cup, like Pharaoh's cupbearer did for him. This was a common practice in ancient times, as the evidence below will reveal.

Pliny, c. AD 70, writes, "As to varieties in respect of size, colour and flavours of the berry [grapes] they are innumerable and they are actually multiplied by the varieties of wine...Democritus, who professed to know all the different kinds of vines in Greece, was alone in thinking it possible for them to be counted, but all other writers have stated that there is a countless and infinite number of varieties; and the truth of this will appear more clearly if we consider the various kinds of wines." [52] Pliny mentions "table grapes," those that "stand carriage well," and those that are good to make raisins..."Some vines which are distinguished for their grapes and not for their wine"...The Raetic grape "ripens in frost. The 'smoke grape,' the 'mouthful' and the tharrupia, which grow on the hills of Thurii, are not picked before there has been a frost." [53]

Clear evidence supports that vines were capable of multiple annual harvests. Varro (c. 36 BC) records, "For the early grapes, and the hybrids, the so-called black, ripen much earlier and so must be gathered sooner; and the part of the plantation and the vineyard which is sunnier should have its vines stripped first." [54] Andrew Dalby reinforces this in his research, "Pepperine wine...where Varro said the vines cropped twice each year...*Apiana* was a pair of varieties, one early, one late." [55]

It was a common practice to squeeze a bunch of grapes by hand directly into a cup and drink that fresh, sweet (fermentation takes away the sweetness), unfermented wine. "Then Pharaoh's cup was in my hand; and I took the grapes and pressed them into Pharaoh's cup, and placed the cup in Pharaoh's hand" [56] Pharaoh, the most powerful man in the world, apparently preferred his wine fresh and unfermented. First century historian Josephus refers to this practice of Pharaoh, and calls this fresh grape juice "wine." A New York Times article referred to the ancient Egyptian pharaohs drinking un-intoxicating beverages. "Nonalcoholic drinks are not new. There is evidence that they existed as far back as ancient Egypt. Nonalcoholic brews were supposedly reserved for the pharaohs, perhaps to allow them to keep a clear head for ruling the land." [57] Even Bacchus, the Roman god of wine is displayed, in sore need of a loin cloth, squeezing grapes by hand into a cup in a stone relief found in the Roman city of Pompeii. [58]

Early church writings referred to pressing grapes into a cup for the Lord's Supper. Dr. Leon C. Field quotes early believers (c. AD 600), orthodox and otherwise, who wrote about pressing grapes directly into the cup at the observance of the Lord's Supper. This would obviously refer to the use of unfermented, non-alcoholic wine used for the Lord's Supper. [59] Cyprian claimed, "Some even who presented no other wine at the sacrament of the Lord's cup but what they pressed out of the cluster of grapes."[60] From the records of *Acts and Martyrdom of Matthew* we discover, "Bring ye also as an offering holy bread, and, having pressed three clusters from the vine into a cup, communicate with me, as the Lord Jesus showed us how to offer up when He rose from the dead on the third day." [61]

Kiddush is a Jewish ceremony and prayer. According to the entry for "wine" in a Jewish Encyclopedia, "One may squeeze the juice of a bunch of grapes into a cup and say the "Ḳiddush" (B. B. 97b)." [62] This shows that squeezing a bunch of grapes directly into a cup was common and accepted by the Jews.

The Rev. Smylie Robson, missionary at Damascus, wrote, "From August to December, bread and grapes are, substantially, the food of the people." [63]

So, for much of the year, grapes could be cut right off the vine and pressed into fresh, sweet, unfermented wine or grape juice. The evidence also shows this was a common practice.

3. Long Shelf Life of Some Grapes

Bunches of grapes could easily be preserved fresh and made available throughout the year. Some will protest their grapes don't keep long, even when placed in the refrigerator. Let me explain.

Long before the birth of Jesus in Bethlehem, people were well aware of the keeping qualities of some grapes. They would hang clusters by their stems in a cool cellar or cave. Such grapes would keep fresh for six months or more. Any old time gardener will tell you some fruits and vegetables are "good keepers;" others are not. A good keeper, at room temperature or placed in a cellar, can remain fresh for months. This was especially well-known in ancient times when such knowledge could mean the difference of having plenty to eat, going hungry, or even starvation.

Characteristics of good keeping grapes include a tough skin and adhering well to the cluster. The cluster would be cut from the vine. Any bad grapes would be clipped, not pulled, from the cluster. Pulling a grape leaves a "brush" that can start a molding, decaying process. Grape clusters were loosely packed in straw, cotton, bran, or hung from the ceiling. Periodically they would be inspected and any bad grapes clipped off. Some were stored in airtight containers. The right varieties of grapes stored in these ways would last fresh for months. Grapes could also be preserved in must or in honey. Good keepers are firm, meaty types with a fairly tough skin. Modern grape varieties that keep well include: Himrod, Interlaken, Lakemont, Seeded Canandaigua, Emperor, Almeria, and Calmeria. [64]

Evidence for preserving fresh bunches of grapes year around is abundant:

The Methodist scholar Leon Fields shows his knowledge of this practice when he states, "Niebuhr says that, 'the Arabs preserve grapes by hanging them up in their cellars, and eat them almost through the whole year.' Dr. Kerr says, 'A friend of mine now in Britain not long since unpacked grapes he had received eleven months previously from the continent, finding them fresh and good.'" Also, "Bernier says grapes were sent from Persia to India, wrapped in cotton, two hundred years ago, and sold there throughout the year." [65]

We can see further from the writings of Pliny that, "The crossed vine has black grapes and makes a wine that does not keep at all long, but its grape keeps a very long time...Some grapes will last all through the winter if the clusters are hung by a string from the ceiling...[and] Grapes are also preserved in must, and so made drunk with their own wine, and some are made sweeter by being placed in must that has been boiled down; but others remain on the parent vine to await the coming of a new generation"... Pliny mentions two kinds of grapes that can be stored in a jar and, "kept in new wine boiled down and must, and properly in after-wine." [66] Three types of wine are mentioned in this last quote by Pliny: new wine boiled down, must, after-wine. The first two un-intoxicating, the third perhaps of a very low alcohol content. All were referred to as wine.

Turning back to Dalby, who researched extensively the ancients, we find a common practice to be describing the kinds of grapes that were good keepers. The next several quotes reflect his reasearch, *"Praecia* or *pretia* included a larger and a smaller variety...these were good for conserving. Columella DA 3.2.23; Pliny NH 14.29." *"Raetica*...also produced the conserved table grapes that Augustus liked...*Aminnia maior* could be conserved as table grapes." *"Duracina* was a group of varieties or category of fruit especially suitable for marketing as table grapes. The name implies firmness of flesh and resilience of skin, and thus suitability for transport to market. Cato DA 7.2 with Dalby ad l.; Columella DA 3.2.1; Pliny NH 14.42; Suetonius, Augustus 76 (quoted at AUGUSTUS)." *"Venucula*...also provided table grapes." [67] This practice was also evident for other crops as seen, "In Athens, 'In the middle of winter you will see chate melons, grapes, fresh fruit...'" [68]

Varro (c. 36 BC) mentions fruits which are hung, thereby preserving them. Some of the fruits mentioned are grapes, apples, and sorbs. "Cato says that the smaller and larger Aminnian grape, and the Apician, are best stored in jars, and that the same grapes keep well also in boiled or plain must; and the best ones for drying are the hard grapes and the Aminnian...At the vintage the careful farmer not only gathers but selects his grapes; he gathers for drinking and selects for eating. So those gathered are carried to the wine-yard, thence to go into the empty jar; those selected are carried to a separate basket, to be placed thence in small pots and thrust into jars filled with wine dregs, while others are plunged into the pond in a jar sealed with pitch, and still others go up to their place in the larder." [69]

This method is even evident in modern times. An 1827 guide tells how to keep grapes and concludes, "In this way grapes may be kept fresh, and perfectly sweet, for 7 or 8 months." [70]

Further, an 1820 Family Receipt (recipe) Book instructs to place a layer of bran in a barrel that can be sealed. Place one layer of grape bunches (that have been gathered in the afternoon of a dry day) on bran. Then continue adding layers of grapes and bran, not allowing grapes to touch each other. Seal barrel so air cannot get in. "Grapes, thus packed, will keep nine or even twelve months." [71] From 1879, "The Pacific Rural Press says: Grapes in April were on exhibition at the store of Strong & Williamson, on Clay Street. The fruit has been preserved in the powdered bark of the sugar pine, and is, in taste and appearance, almost as fresh as the day it was taken from the vine. They were received from George Geissendefer, of Placerville. The fact that grapes can be preserved in this way may be suggestive to some of our vine growers. Certainly there could be money made by supplying the trade with fresh grapes at this time of the year." [72]

Current expressions of this practice are recorded in the modern-day book, *Root Cellaring*, which has a chapter devoted to "Good Keepers." The book discusses the fruits and vegetables that have proven themselves to keep well after harvest. [73] Ancient people of Bible times and the Mediterranean world knew well these qualities, for often their lives depended on it. More recently we see, "Grapes are put in large deep baskets and stacked in a cool place for storage. The grapes that can be stored are the white and red varieties; the others have short lives. Red grapes are also stored by hanging them on the ceiling. Black grapes are used in making grape molasses (pekmez), and are dried for winter consumption. White grapes are used for making pestil (mulberry leather) and orcik." [74]

From this information it is easy to see how fresh grapes were accessible months after harvest. This information refutes the argument that Passover wine had to be fermented since Passover was in the spring, long after the fall grape harvest. This information also refutes the argument that the only choice people had before electricity, refrigeration, and pasteurization was to partake of fermented wine. These grapes, preserved fresh throughout the year, could be pressed into a cup at any time. Whether it was for observance of the Lord Supper, Passover, or just for refreshing sweet wine, unfermented juice could be produced at any time.

4. Reconstituted Dried Grapes or Raisins.

Drying is one of the oldest methods of preserving food. It is still used today. Grapes were dried in the sun and covered at night to protect them from the dew. Grapes were also dried through heat and smoke. The dried grapes or raisins were re-hydrated by soaking or boiling and pressed into wine. This raisin wine was sometimes called passum. Raisin wine was used in ancient times, and in some areas still used today.

The Talmud (ancient Jewish writings) refers to raisin wine. Polybius, Greek historian c. 100 BC, spoke of un-intoxicating raisin wine. A medieval Arabian writer refers to raisin wine for the Lord's Supper. Modern day Jews refer to raisin wine.

Someone may object, "But I found a reference to raisin wine that was clearly intoxicating." You can make any drink on the face of the earth, into an intoxicating drink. But the point here is that raisins have been, and can be, made into a fresh un-intoxicating wine at any time of the year in any place on earth. Yes, it can also be made alcoholic. A soft drink can be made alcoholic. But raisin wine is a plain example of yet another way the ancients could, and did, make and preserve unfermented wine. Consider the following evidence.

The Greek Historian Polybius, c. 100 BC, demonstrates this was a generally accepted practice when he writes, "Among the Romans women are forbidden to drink wine; and they drink what is called passum, which is made from raisins, and tastes very like the sweet wine of Aegosthena or Crete. This is what they drink to quench their thirst." [75] That this raisin wine tasted like sweet wine and quenched their thirst is further evidence that it was nonalcoholic wine.

Raisin wine is mentioned often in ancient literature. Hippocrates and Celsus regularly refer to raisin wine. "The Talmud mentions a wine made from raisins (T.B. Baba Bathra 97b)." [76] It is accepted in modern kosher literature, "However, raisin wine, that is, when water is poured upon the raisins, is considered as true wine." [77]

Again Pliny makes great contribution to this reality when he says, "Psythium and melampsythium are varieties of raisin-wine which have the peculiar flavour of the grape, and not that of wine." [78] This raisin wine is called wine. This wine, however, tastes like unfermented wine or grape juice, rather than alcoholic wine. This raisin wine is clearly nonalcoholic. He mentions, "table grapes," those that "stand carriage well," and those that are good to make raisins." [79] Further he adds clarity to the unfermented nature of this wine when he writes, "Must prepared from raisins of the sun has a more pleasant flavour and is productive of a less degree of oppression to the head." [80] Raisins were popular in ancient times and not only used for wine, but as a food ingredient, sweetening, and eating out of hand. Smoked raisins were a Roman favorite and enjoyed by Augustus and Tiberius. [81]

The sample from the ancient practitioner of medicine, Celsus' medical prescriptions include the following phrases on raisin wine: "compounded with raisin wine." "which are taken up in honey or in raisin wine." "gradually adding raisin wine." "mixed an equal quantity of raisin wine." "boiled in raisin wine." [82]

The prevalence of passum is evident in the following quotes, "The ancients frequently speak of the Cretan wines (Ael. VH 12.31; Athen.10.440; Plin.Nat. 14.9). Among these the 'passum,' or raisin wine, was the most highly prized (Mart. 13.106; Juv. 14.270)." [83] Note a nonalcoholic wine was most highly prized! "At Lachish, an inland city of Judah, a jar was discovered with an inscription in early Semitic reading: 'Wine made from black raisins.' Drying the grapes on the vine or spreading them out on mats in the sun concentrated the sugar, and very sweet wines could be made by this method." [84] "The raisin was a crucial part of Hittite life, whether as a dry ration on a military campaign or as a special offering for the king or the gods. These people had perfected raisin wine, as one tablet puts it: '[Behold] the raisin. Just as it holds its wine in [its] heart…'" Any "wine" in a raisin is unfermented wine. McGovern goes on to say the Hittite texts are replete with references of "sweet wines." [85]

Bersalibi, an Arabian believer in medieval times, stated in a tract on the Eucharist, "When good wine is not to be obtained, the juice of the grapes may be taken; or the liquor expressed from dried grapes or raisins." [86] It should be remembered that "liquor," like "wine," can refer to either an alcoholic or non-alcoholic liquid. George Whitefield Samson points out that this reference to "good wine" denotes unfermented wine and that Pliny and others referred to unfermented wine in this way. In the 1500s Bishop Osorius said of Christians in the south of India, "They use wine prepared from dried grapes in their sacrifices." [87] Thevenot said of old time Christians dwelling in the Jordon River valley, "As for the wine of their consecration, they make use of wine drawn from dried grapes steeped in water, which they express." [88]

From the writings of Passover observing Jews we find, "'Unfermented liquor or wine free from alcoholic substances…is used to the present day at the Passover, (it is) the wine over which the blessing is said,' wrote the New York politician and journalist, Mordecai M. Noah, in a letter to the editor of the New York Evening Star for the Country in 1838." [89] "A. J. Gordon sought out the practice of a distinguished rabbi of his day, namely Rabbi Simeon Reinstein of Boston, who wrote to Dr. Gordon: 'The wine which we use at Passover is made from pure raisins, and is unfermented.'" [90] In New York City, "The Passover wine is prepared from crushed raisins or dried grapes, steeped in water, pressed and made into a sweet but unfermented wine." [91]

Jews are all over the map on the issue of drinking and the use of fermented or unfermented wine. A significant number of Jews down through the years, however, have insisted on unfermented (and unleavened) wine for the Passover. Often they would make this unleavened nonalcoholic raisin wine for Passover. This is not to say that all Jews use nonalcoholic wine for Passover, but through the years, many have certainly done so.

Even cookbooks from antiquity to modernity speak of this sweet wine: From Rome c. AD 400, "Crush pepper, lovage, with the broth, add a little raisin wine to sweeten. Cook it, thicken with roux, immerse the balls in the sauce and serve." [92] This is evidently un-intoxicating wine because it is raisin wine, and because it sweetens. From an AD 1947 Jewish cookbook, "Cut 1 lb. raisins in halves, put in a saucepan with 3 pints cold water and simmer *very gently* until the water is reduced by one third. This should take several hours. When cold, strain through muslin. If additional sweetness and flavouring are required, add a little sugar and cinnamon before boiling." [93] Not only does this give a modern day nonalcoholic raisin wine recipe, but it calls a clearly nonalcoholic raisin drink, "wine." Joan Nathan, a graduate of the University of Michigan and Harvard University and has lived and worked in Jerusalem and has written food articles for the *Washington Post, New York Times, Food Arts, Gourmet,* and *B'nai B'rith International Jewish Monthly.* In *Jewish Cooking* she states, "Although raisin wine is not the same as wine made from grapes, it was one of the early methods of winemaking used by Jews throughout the world since the biblical period." [94]

One last quote by Mordecai M. Noah on raisin wine: "...take a gallon demijohn, or stone jug; pick three or four pounds of bloom raisins, break off the stems; put the raisins into the demijohn, and fill it with water. Tie a rag over the mouth, and place the demijohn near the fire, or on one side of the fire-place, to keep it warm. In about a week it will be fit for use, making a pure, pleasant, and sweet wine, free from alcohol. It may last from Sunday to Sunday without getting sour or tart; but it is easy to make a small quantity of wine for each time it is to be used. This is the wine we use on the nights of Passover." [95] Again notice this raisin drink, free from alcohol, is called wine. Unfermented raisin wine has been often used in ancient and modern times.

5. Sealed Wine Containers.

We have three sets of ancient instructions for an additional way to keep sweet new wine or must. This was done by sealing airtight new wine (grape juice) and immersing it in cold water. Notice the use of the terms, "always sweet," and "always must."

Almost two centuries before the birth of Christ, Roman writer Cato said, "If you wish to keep grape juice [must, or new wine] through the whole year, put the grape juice in an amphora [flask, container for wine], seal the stopper with pitch, and sink in the pond. Take it out after thirty days; it will remain sweet the whole year." [96]

During New Testament times Roman writer Columella said, "That must [new, sweet, unfermented wine] may remain always as sweet as though it were fresh, do as follows. Before the grape-skins are put under the press, take from the vat some of the freshest possible must and put it in a new wine-jar; then daub it over and cover it carefully with pitch, that thus no water may be able to get in. then sink the whole flagon in a pool of cold, fresh water so that no part of it is above the surface. Then after forty days take it out of the water. The must will then keep sweet for as much as a year." [97] Keeping sweet signifies un-intoxicating wine that has not lost its sugar content through alcoholic fermentation.

Pliny, c. AD 70, agrees, "The liquor to which the Greeks give the name of 'aigleucos,' [always sweet] is of middle quality, between the sirops and what is properly called wine; with us it is called 'semper mustum' [always must]. It is only made by using great precaution, and taking care that the must does not ferment; such being the state of the must in its transformation into wine. To attain this object, the must is taken from the vat and put into casks, which are immediately plunged into water, and there left to remain until the winter solstice is past, and frosty weather has made its appearance. There is another kind, again, of natural aigleucos, which is known in the province of Narbonensis by the name of 'dulce' [sweet], and more particularly in the district of the Vocontii. In order to make it, they keep the grape hanging on the tree for a considerable time, taking care to twist the stalk. Some, again, make an incision in the bearing shoot, as deep as the pith, while others leave the grapes to dry on tiles. The only grape, however, that is used in these various processes is that of the vine known as the 'helvennaca.'" [98]

As stated elsewhere, sometimes ancients wanted to call wine, only that which is fermented. But often that was not the case and sweet unfermented grape juice was called wine (see chapter 3). Regardless, ancient writers said the methods above will keep new wine or must from fermenting.

6. Lactic Fermentation of Wine

I have often used the term unfermented wine to refer to that which is not intoxicating. I continue to use that term, although technically, there is an exception. The ancients actually had fermented wine that would not intoxicate.

How can that be? There is more than one kind of fermentation. We usually think of alcoholic fermentation. This is when yeast devours sugar and produces alcohol (and a gas, carbon dioxide). There is also acetic fermentation that converts alcoholic wine into nonalcoholic sour wine or vinegar. But there is another type of fermentation the ancients knew well.

In study of ancient wine I continued to be confronted with salt as a common ingredient to many wines. Modern-day wine making does not include salt. I understood the importance of salt in preserving things like fish and the old fashioned salt-cured country hams. But why was salt added to wine? In thumbing through a *Lehman's Non-Electric Catalogue*, I came across the book, *Preserving Food without Freezing or Canning*. The book was ordered, and it was the first I read that explained the method of salt preserving vegetables and consequently, non-alcoholic wine. The process is today called lactic fermentation.

Lactic fermentation was used throughout the world until very recent times. Now canning with pasteurization and freezing have pretty well taken its place. While most know nothing of this type of food preservation, lactic fermentation is still practiced by many, especially in other countries. Some praise this method for its health benefits as well. The reason is that through lactic fermentation you are ingesting a live culture, something like the more familiar yogurt, buttermilk, or un-pasteurized vinegar (with the "mother"). [99] Many say it helps aid the digestive system.

In our consideration of the Bible and wine, it should be noted that the great majority of modern day scholars know little to nothing about ancient preservation of non-alcoholic wine, and even fewer would know about lactic fermentation. Ever heard of a seminary course on ancient food preservation? Yet this is a key to understanding the common wine of Bible days.

Sauerkraut is made by this method, though nowadays the process usually ends in pasteurization (heating the food to a certain temperature that would kill any live culture, good or bad). Pickles used to be made this way. And wine was commonly made this way. Ancients also commonly preserved food by lactic fermentation in brine, or salt water. [100]

How was ancient wine made? We find it in their literature and especially their wine recipes. We have actual recipes of ancient wine. Some of those ancient recipes prepared and preserved wine by lactic fermentation. Cato (c. 170 BC) and Pliny (c. AD 70) clearly give these recipes. Wine made from these instructions would have been preserved by lactic fermentation and would have had no alcohol, or very small amounts of alcohol. Remember, even fresh orange juice has small amounts of alcohol. And the human body naturally produces small amounts of alcohol.

Ancient people were very knowledgeable and intelligent. But they had no idea of the good and bad microorganisms. They just knew some things worked and some did not. Some things were healthful and some weren't. Some practices successfully preserved food; some did not.

Beatrice Trum Hunter teaches, "Through trial and error, and observation, man learned that when natural fermentations occurred, and salt was added, there were definite changes in the odor, appearance and taste of foods, but they remained wholesome." [101] Salt is central to many of the ancient wines. Preserving vegetables with lactic fermentation is explained in the following quote. "Salt is a key: A good rule of thumb is about 1 ½ percent salt by weight of vegetables, which generally translates into 2 to 3 tablespoons of salt per quart. Lactic microbial organisms - similar to those that curdle milk - develop spontaneously and convert the natural sugars of the vegetable into lactic acid. This environment rapidly acidifies, to the point that it becomes impossible for bacteria responsible for food spoilage to multiply. Vegetables preserved this way will keep in a cool place, such as a cellar, for many months." [102]

Now, see some of the following ancient wine recipes. These common Greek and Roman wines would have been non-alcoholic, or would have had a very low alcohol content. These recipes use the word wine for what would have been nonalcoholic beverages. Bear in mind this was the world in which the Jews and early Christians lived.

Marcus Porcius Cato (aka Cato, Cato the Elder) lived from 234-149 BC. Cato was a Roman military leader, writer, and agriculturalist. He lived during the intertestamental period, or the time between the close of the Old Testament (c. 400 BC) and the birth of Jesus in Bethlehem (c. 5 BC). His book, *On Agriculture* (De Agri Cultura), is an early farmer's notebook. Following are some of his non-alcoholic wine recipes.

"Directions for making Greek wine: Gather carefully well-ripened Apician grapes, and add to the culleus of must two quadrantals of old sea-water, or a modius of pure salt. If the latter is used, suspend it in a basket and let it dissolve in the must [new wine]. If you wish to make a straw-coloured wine, take equal parts of yellow and Apician wine and add a thirtieth part of concentrated must to any kind of blended wine." [103]

"If your place is far from the sea, you may use this recipe for Greek wine: Pour 20 quadrantals of must into a copper or lead boiler and heat. As soon as the wine boils, remove the fire; and when the wine has cooled, pour into a jar holding 40 quadrantals. Pour 1 modius of salt and 1 quadrantal of fresh water into a separate vessel, and let a brine be made; and when the brine is made pour it into the jar. Pound rush and calamus in a mortar to make a sufficient quantity, and pour 1 sextarius into the jar to give it an odour. Thirty days later seal the jar, and rack off into amphorae in the spring. Let it stand for two years in the sun, then bring it under cover. This wine will not be inferior to the Coan." [104]

"Recipe for Coan wine: Take sea-water at a distance from the shore, where fresh water does not come, when the sea is calm and no wind is blowing, seventy days before vintage. After taking it from the sea, pour into a jar, filling it not fully but to within five quadrantals of the top. Cover the jar, leaving space for air, and thirty days later pour it slowly and carefully into another jar, leaving the sediment in the bottom. Twenty days later pour in the same way into a third jar, and leave until vintage. Allow the grapes from which you intend to make the Coan wine to remain on the vine, let them ripen thoroughly, and pick them when they have dried after a rain. Place them in the sun for two days, or in the open for three days, unless it is raining, in which case put them under cover in baskets; clear out any berries which have rotted. Then take the above-mentioned sea-water and pour 10 quadrantals into a jar holding 50; then pick the berries of ordinary grapes from the stem into the jar until you have filled it. Press the berries with the hand so that they may soak in the sea water. When the jar is full, cover it, leaving space for air, and three days later remove the grapes from the jar, tread out in the pressing room, and store the wine in jars which have been washed clean and dried." [105]

Pliny was a Roman writer who lived during New Testament times. Pliny mentions wine called "life" that had a large amount of sea-water mixed in. He relates how some wine is called "sea-flavoured wine" and "sea-treated." Some wines were saltier than others. The following are samples of his recipes.

A sweet honey wine "is made from must [new wine], in the proportion of thirty pints of must of a dry quality to six pints of honey and a cup of salt, this mixture being brought to a boil; this produces a dry-flavoured liquor."[106] If the must had any alcohol content, it is excluded when this mixture is boiled. The salt, preserves it from converting to alcohol. As previously mentioned, in lactic fermentation the wine ferments, but not into alcohol.

"Boil tender sprigs of myrtle with the leaves on in salted must, and after pounding them boil down one pound of the mixture in 2 ¼ gallons of must until only 1 ½ gallons are left. The beverage made by the same process from the berries of the wild myrtle is called myrtle wine;

this stains the hands." [107] Three factors demonstrate this is non-alcoholic wine. First, it is boiled. Second, it is boiled into a thick consistency that would retard alcoholic fermentation. Third, the salt content retards alcohol fermentation and promotes lactic fermentation.

"Wine is made from the root of asparagus." Pliny gives the recipe: asparagus root, cunila, wild marjoram, parsley-seed, southernwood, wild mint, rue, catmint, wild thyme, horehound, must, boiled down must [grape-juice, new wine], and sea-water. [108] This concoction should have been able to cure anything! In this recipe we also see not only the preservative salt, but also the preservative marjoram.

Columella, a contemporary of Pliny writes, "Since some people - and indeed almost all the Greeks - preserve must with salt or sea-water." Columella also advises, "Also salt and spices, which he has been accustomed to use in the preservation of wine, ought to be stored up in good time beforehand." [109] Athenaeus, from the third century AD writes, "Wines which are more carefully treated with sea water do not cause headache." [110] Other quotes from Athenaeus on salt and wine, and unintoxicating wine, are given in the next chapter.

Last, let me give a modern day lactic fermentation wine recipe. Note the similarity to wine recipes of 2,000 years ago.

"Makes five to six quarts. About 16 pounds of organic red grapes; ½ cup whey; 1 tablespoon sea salt. This delicious and refreshing drink is an excellent substitute for wine, containing all the nutrients of grapes found in wine, including many enzymes, but none (or at least very little) of the alcohol. Remove grapes from stems, wash well and pass through a juicer. Place liquid in a large bowl with salt and whey and stir well. Cover with a cloth and leave at room temperature for three days. Skim off any scum that may rise to the top and strain juice through a strainer lined with a tea towel. Store 'grape cooler' in airtight glass containers in refrigerator. Delicious flavors will develop over time. Best served diluted - half water, half grape juice. The sediment will fall to the bottom of the container and should remain there if the grape cooler is poured out carefully. However, you may also filter the cooler again by pouring it through a strainer lined with a tea towel." [111]

Interestingly this recipe mentions that flavors will develop over time. So non-alcoholic wine can also age and develop flavors. Also significant is that this recipe says it is best served diluted with water. The ancients almost always diluted their wine with water. Just as in alcoholic fermentation, in lactic fermentation the wine ferments, and the wine leaves sediment, or lees.

This modern day recipe says it will produce little or no alcohol in the drink. This recipe calls for very little salt. With a little more salt, any small alcohol content would be even less or non-existent. Whey is present in the air and some do not add it, but it can help jump start the process of lactic fermentation. Whey is mentioned by Columella in the preservation of food. Whey is the liquid that strains out of live culture yogurt.

So among many ancient methods of preserving non-alcoholic wine, one of those common methods was with salt and lactic fermentation. As previously noted, this is not meant to deny the ancients had alcoholic wine; they obviously did. But they also had alcohol free wine.

7. Boiling Out Alcohol in Wine

Ancient people had no real knowledge of microorganisms and germs. They did not have a word for alcohol and did not understand that boiling alcoholic wine boiled away the alcohol. But they often employed boiling with both wine and water.

They did not understand the details, but they knew that boiling somehow made a liquid clean and safe.

They would boil new wine; they would boil old wine. Boiling wine is often mentioned in ancient literature. Alcohol boils away at a lower temperature than water. Even if the wine were fermented, boiling would take the alcohol away. With new expressed wine, boiling would kill the yeast or ferment and kill any other impurities.

8. Sulfur Preserved Wine

Look at the label on much of your un-intoxicating food and drink today. Sulfur dioxide is often on the list of ingredients. Sulfur dioxide and sulfites are used as preservatives. Sulfur prevents browning in alcoholic beverages, fruit juices, soft drinks, fruits and vegetables. Sulfur prevents yeast and bacterial growth. Of course this is a modern day food preservative unknown until modern times. Right? Wrong. Pliny and other ancient writers mentioned the use of sulfur to preserve wine.

One pro-drinking writer argues that sulfur was indeed used as a wine preservative, but only to preserve fermented wine from going bad and turning into vinegar. He is half right and half wrong. Sulfur is indeed used today by winemakers to prevent alcoholic wine from turning to vinegar. But it also has another use.

Makers of alcoholic wine today sometimes buy must and the ingredients to make their own wine at home. The producers of these kits ship them a five gallon container of must. They do not want the must to begin to ferment before the purchaser gets it in the mail. Therefore they often treat the must with low levels of sulfur to prevent the must or grape juice from fermenting too early.

This practice, however, sometimes presents another problem. The home winemaker receives his five gallons of must, adds the yeast and sugar. Occasionally even after this, the wine does not begin the fermentation process. This is because even with the yeast and sugar the small amount of sulfur holds back fermentation. Winemakers list this as one of the problems sometimes experienced with these winemaking kits. So sulfur can keep fermented wine from turning to vinegar. It can also prevent sweet unfermented wine from fermenting in the first place.

The pro drinking authority also says the ancients would have had to add too much, a lethal dose, of sulfur to new wine or grape juice to keep it from fermenting. Wrong again. It was a common practice for ancients to boil their wine right after pressing it from the grapes. That would cleanse it and kill the yeast in it. Then relatively small amounts of sulfur, as mentioned above, could retard fermentation.

9. Other Preservatives in Wine

Pliny lists the following preserving and additive ingredients sometimes used in wine: gypsum, lime, potter's earth, marble dust, salt, sea-water, resinous pitch, season must with resin, ashes, sulfur. [112] Marble dust (calcium carbonate) has various uses today, including use as a preservative. Among others, it is used in chewing gum, toothpaste, and food.

Elsewhere ancient writers mention the preservative marjoram being used in wine. Modern day scientists know that marjoram has preservative qualities.

Additionally, olive oil and resin were used to make containers and contents airtight.

Finally, filtering was claimed to break the strength of wine, and could filter out alcohol content. [113] Alcohol molecules are larger. This is the process used by makers of dealcoholized wine. Finally, new wine does not begin to ferment if the temperature is too low or two high.

Nonalcoholic wine was common in the ancient world. It was common among the Egyptians, Mesopotamians, Hittites, Greeks, Romans, and Israelites. As this chapter documents, ancient people had multiple ways of preserving nonalcoholic wine without modern day pasteurization, electricity, and refrigeration. These methods were widely practiced and provided unfermented wine throughout the year. So don't let anyone, scholar or otherwise, tell you it was impossible to keep wine from fermenting in ancient times. Don't let them tell you that nonalcoholic wine was unavailable during most of the year. Such statements reveal a fundamental lack of understanding of common ancient practices. Just as today, they had a choice of drinking alcoholic or nonalcoholic beverages.

One last, dubious method of producing nonalcoholic wine, "Hence Pope Julius I criticized some who 'keep throughout the year a linen cloth steeped in must, and at the time of sacrifice wash a part of it with water, and so make the offering.'" [114] I would have to agree with the Pope on this one. But then, this method would certainly produce un-intoxicating, and weak, wine!

CHAPTER THREE:
Evidence of Non-Alcoholic Wine in Ancient Times

Why does it matter whether they had non-alcoholic wine in ancient times? The Bible did not arise in a vacuum. While the Bible is the divinely inspired, inerrant Word of the Living God, it used the common language of the day. The people of Bible times had many similar practices with the people of the surrounding countries. To learn more about the world in which biblical people lived is to learn more about the culture of the Israelites. Just as the wine in the Bible was generic and included both alcoholic and non-alcoholic wine, we find this was also true in the surrounding ancient world.

Bible scholars are interested in the customs of the Bible and customs of the indigenous people and surrounding people of Bible lands. Scholars are also interested in extra-biblical use of language from ancient times because that provides a fuller understanding of language that is used in the Bible.

As previously mentioned, ancient Israel was at the crossroads of the world. Trade routes passed through Israel from Europe, Asia, and Africa. Ships came to and from Israel from the Mediterranean world. Israel was knowledgeable about and influenced by Egypt, Greece, Rome and other countries. So it helps our understanding of Scripture to see how people referred to wine in ancient times. Was wine only used of that which intoxicates? The answer is plainly no. Surrounding nations referred to wine, as the Bible does, as that which is alcoholic and that which is not alcoholic. We will see from the writings of Aristotle, Hippocrates, Homer, Athenaeus, Columella, Pliny, Plutarch and many others from antiquity the reality of the general use of the term wine to designate nonalcoholic wine.

Aristotle, c. 350 BC

Aristotle (384 BC - 322 BC) was a Greek philosopher. A student of Plato, he taught Alexander the Great. Aristotle's writings have been very influential in Western culture. He lived just after the close of the Old Testament. His statements clearly show that nonalcoholic grape juice was considered wine.

"Water on the other hand does not give off fumes, but vapour. Sweet wine does give off fumes, for it contains fat and behaves like oil. It does not solidify under the influence of cold and it is apt to burn. Really it is not wine at all in spite of its name: for it does not taste like wine and consequently does not inebriate as ordinary wine does. It contains but little fumigable stuff and consequently is inflammable." [115] Aristotle seems to want to say only intoxicating wine is wine, but then he points out other kinds of wine and the different properties of different wine. He wants to say sweet un-intoxicating wine is not wine at all, but he adds that it is after all, named wine. In other words, sweet un-intoxicating wine is, in fact, wine. Notice that he says "Sweet wine...does not inebriate" or intoxicate. Also notice this sweet wine does not taste like wine. Nonalcoholic sweet wine is distinctively different in taste from alcoholic wine.

"Those liquids which are thickened by heat are a mixture. (Wine is a liquid which raises a difficulty: for it is both liable to evaporation and it also thickens; for instance new wine does. The reason is that the word 'wine' is ambiguous and different 'wines' behave in different ways. New wine is more earthy than old, and for this reason it is more apt to be thickened by heat and less apt to be congealed by cold. For it contains much heat and a great proportion of earth, as in Arcadia, where it is so dried up in its skins by the smoke that you scrape it to drink. If all wine has some sediment in it then it will belong to earth or to water according to the quantity of the sediment it possesses.) The liquids that are thickened by cold are of the nature of earth; those that are thickened either by heat or by cold consist of more than one element, like oil and honey, and 'sweet wine'." [116] Two very important points are found in this quote by Aristotle. First, he says the word "wine" is ambiguous, and different wines behave in different ways. That is exactly the contention of this book. There are both alcoholic and non-alcoholic wines, and their properties are very different. Second, Aristotle says the new wine of Arcadia "is so dried up in its skins by the smoke that you scrape it to drink." This is a clear reference to non-alcoholic wine.

A slightly different translation of Aristotle, "The reason is that there is more than one kind of liquid called wine, and different kinds behave in different ways." [117] Pick either translation, both prove the point that wine refers to different kinds of wine with very different characteristics.

Elsewhere Aristotle states, "Gold, then, and silver and copper and tin and lead and glass and many nameless stones are of water: for they are all melted by heat. Of water, too, are some wines and urine and vinegar and lye and whey and serum: for they are all congealed by cold. In iron, horn, nails, bones, sinews, wood, hair, leaves, bark, earth preponderates." [118] Notice Aristotle says "some wines" have these properties. Alcoholic wines have certain properties; non-alcoholic wines have different properties. Similarly, "Now those things that are not thickened by cold, but solidified, belong rather to water, e.g. wine, urine, vinegar, lye, whey." [119] Likewise, "Thus any mixed drink is called oinos, 'wine'. Hence Ganymede is said 'to pour the wine to Zeus,' though the gods do not drink wine. So too workers in iron are called chalkeas, or 'workers in bronze.' This, however, may also be taken as a metaphor." [120] Aristotle points out the generic, general usage of the word wine. He even uses the Greek word for wine, oinos. This is one of the words used for wine in the New Testament and the Septuagint (ancient Greek translation of the Hebrew Old Testament). According to Aristotle, the usage of oinos, or wine, is so general that any mixed drink can be called wine. This is in glaring opposition to those in our modern day who proclaim oinos always refers to intoxicating wine.

Aristotle very significantly refers to "must" as a kind of wine. Must is new, unfermented wine. "There is a kind of wine, for instance, which both solidifies and thickens by boiling - I mean, must." [121] Some dictionaries define must as "new wine." Aristotle makes distinction between sweet wine and dry wine when speaking of reasons for preference, "For the 'desire of X' may mean the desire of it as an end (e.g. the desire of health) or as a means to an end (e.g. the desire of being doctored), or as a thing desired accidentally, as, in the case of wine, the sweet-toothed person desires it not because it is wine but because it is sweet. For essentially he desires the sweet, and only accidentally the wine: for if it be dry, he no longer desires it. His desire for it is therefore accidental. This rule is useful in dealing with relative terms: for cases of this kind are generally cases of relative terms." [122]

Some today strangely feel that a man will choose alcoholic wine every time. Aristotle refutes this notion above, and by saying, "Just as a man chose his wine because it was sweet". [123] Notice the word "wine" stands alone, and it is used to refer to sweet wine, nonalcoholic wine. One final quote from Aristotle dealing with diluted wine, he states, "Wine and water is called 'wine,'" [124] This is one of many instances referring to the common practice of mixing wine with water.

Hippocrates

Hippocrates (c. 460 BC - 370) was a Greek medical doctor. He is considered the father of the medical profession. His writings have been very influential through the centuries. Hippocrates lived during the closing days of the Old Testament. A part of the original *Hippocratic Oath* says,

"I will prescribe regimens for the good of my patients according to my ability and my judgment and never do harm to anyone. I will not give a lethal drug to anyone if I am asked, nor will I advise such a plan; and similarly I will not give a woman a pessary to cause an abortion."

In *On Ulcers*, c. 400 BC, Hippocrates gives several recipes for medicinal formulas. He mentions taking wine, boiling it, and after it is boiled he continues to call it wine. Whether the wine was initially alcoholic or not, it was certainly non-alcoholic wine after being boiled. In the same book, Hippocrates also refers to "sweet wine," then later refers to that same liquid simply as "wine." This is pointed out because some social drinkers have contended that at least in ancient times non-alcoholic wine was never referred to as just "wine," but instead always had a qualifier. Well, sometimes it did; sometimes it did not.

While the ancients did not understand the details of the substance of alcohol, they knew some wine affected the head and some did not. While they did not know the details of boiling and what it did to alcohol and microorganisms, they did know that boiling water or wine would purify it.

"One must determine by such marks as these, when sweet, strong, and dark wine, hydromel, water and oxymel, should be given in acute diseases. Wherefore the sweet affects the head less than the strong, attacks the brain less..." [125] Notice sweet wine affects the head less, attacks the brain less. That is just another way of saying sweet wine was nonalcoholic.

Homer

Homer, in about 800 BC said, "Wait till I can bring you wine that you may make offering to Jove and to the other immortals, and may then drink and be refreshed. Wine gives a man fresh strength when he is wearied, as you now are with fighting on behalf of your kinsmen."

[126] Alcoholic wine has lost its sugar content and is a depressant; not likely to give a fighting man fresh strength when he is wearied. Non-alcoholic wine or grape juice is filled with sugar and would refresh and strengthen a wearied soldier. While the ancients would not have understood all the details, they would know what strengthened a weary man, and what would not. This quote is a likely reference to sweet, un-intoxicating wine.

Athenaeus

Athenaeus lived in Egypt and wrote the *Deipnosophistae*, or "*Sophists at Dinner.*" He quotes many previous ancient writers and says much about wine and dining in his day and before. It was published about AD 220 or 230, some say right after Athenaeus' death. Athenaeus has much to say about wine.

"For wine is sweet when sea water is poured into it." [127] "Wines which are more carefully treated with sea water do not cause headache; they loosen the bowels, excite the stomach, cause inflations, and assist digestion. Examples are the Myndian and the Halicarnassian. The Cynic Menippus, at any rate, calls Myndus 'salt-water drinker.' The wine of Cos also is very highly treated with sea water." [128] "The Privernian [wine] also can be used then, being thinner than that of Rhenium [wine] and not at all likely to go to the head." [129] Again, this indicates no fermentation. "The Statan is one of the best kinds [of wine], resembling the Falernian, but lighter, and innocuous." [130] Innocuous means harmless, innocent, safe, mild. Notice this harmless wine resembled other popular wine. "The Mitylenaeans call the sweet wine of their country prodromus; others say protropus." [131] "Philyllius says: 'I will furnish Lesbian, mellow Chian, Thasian, Bibline, and Mendaean [wines], and nobody will have a headache.'" [132] Lesbian wine gets its name from its place of origin, the Greek Isle of Lesbos just off the coast of modern day Turkey. The term has nothing to do with the modern day use of the word. Modern day residents have strongly protested the homosexual references to their name. All the names in this quote refer to different types of wine that apparently did not produce intoxication or hangover. This indicates how common un-intoxicating wine was in the ancient world. Another reference to wine from Lesbos is, "But we find in the island of Lesbos the protropum wine," [133]

Athenaeus continues, "But the sweet varieties, both of white and yellow wines, are the most nutritious. For sweet wine smooths the tract through which it passes, and by thickening the humours more, tends to incommode [inconvenience, trouble] the head less." [134] Sweet wine usually is referring to un-intoxicating wine. Notice this ancient quote says sweet wine is most nutritious and troubles the head less. That is also true of unfermented wine or grape juice today. "Chian wine promotes digestion, is nourishing, produces good blood, is very mild, and is satisfying in its rich quality." [135] "Nourishing" and "very mild" refer to un-intoxicating qualities. Unfermented wine has a much higher sugar content and is filled with vitamins; the process of fermentation eats up the sugar content, thereby making it less nourishing. The fact that alcohol is a depressant also diminishes its nourishing quality. "The so-called Adriatic has a pleasant odour, is easily assimilated, and altogether innocuous." [136] This Adriatic wine was said to be completely harmless. Can that be said of the common wine of today? "The wine of the Thebaid, and especially the wine from the city of the Copts, is so thin and assimilable, so easily digested, that it may be given even to fever patients without injury." [137] This is another wine that could be used without injury.

"The Mareotan wine — also called Alexandreotic...is excellent; it is white and pleasant, fragrant, easily assimilated, thin, does not go to the head and is diuretic." [138] The ancients did not have a word for alcohol and did not understand the details of the substance of alcohol. Today we would simply say Mareotan Wine is non-alcoholic, in ancient times they explained it by saying it does not go to the head. "Now sweet wines do not make the head heavy, as Hippocrates says in his book *On Diet*, which some entitle, *The Book on Sharp Pains*; others, *The Book on Barleywater*; and others, *The Book against the Cnidian Theories*. His words are: 'Sweet wine is less calculated to make the head heavy, and it takes less hold of the mind, and passes through the bowels easier than other wine.'" [139]

Athenaeus demonstrates his understanding of two basic kinds of wine. "And Alcaeus says - 'Wine sometimes than honey sweeter, sometimes more than nettles bitter.'" [140] Ancient nonalcoholic wine was sweet; alcoholic wine was bitter. Athenaeus here clearly references both kinds of wine. "And if any one thinks it too much trouble to live on this system, let him take sweet wine, either mixed with water or warmed, especially that which is called πρότροπος, the sweet Lesbian wine, as being very good for the stomach." [141] "At the time of festivals, he went about, and took wine (oinos) from the fields." 141 One does not take alcoholic, but nonalcoholic wine from the fields. You have to later work at it to make it alcoholic. This is a reference to wine as being the un-intoxicating wine in the fields, still on the vine, and still in the grape. This quote of Athenaeus is reminiscent of Judges 9:13; Isaiah 65:8; Jeremiah 40:10, 12.

Columella

Columella was a Roman who lived from 4 BC to about AD 70. He was a contemporary of Pliny and, of course, lived in the days of the New Testament portion of the Bible. He wrote extensively on agriculture.

"Wine which you wish to have rather sweet you will have to preserve on the day after you take it out of the vat, but, if you want it rather harsh in flavour, on the fifth day." [142] This is a very important quote when it comes to the question of whether or not they had nonalcoholic wine in ancient times and in Bible times. This quote makes it obvious they did. Sweet wine would need to be preserved rather soon, before alcoholic fermentation. The preservative he is speaking of here is either salt, or sea-water. Notice that in this short sentence Columella gives a recipe for non-alcoholic wine, and also for alcoholic wine. He tells how to make sweet wine or harsh wine. Some winemakers, however, might even wonder whether five days would allow it time for full fermentation, indicating that often even their fermented wine was relatively low in alcohol. Preservatives are used for both kinds of wine: for sweet wine to keep it from losing its sweetness through alcoholic fermentation, thereby preserving it in an un-intoxicating state. And for harsh wine to keep it from further fermenting into acetic acid, sour wine, or vinegar. Preservatives obviously work on both kinds of wine. This is also a clear reference to the fact that sweet wine was normally nonalcoholic wine. Preserving wine as soon as it comes out of the wine-press leaves it sweet because the alcoholic fermentation has not destroyed the sugar content. Last, note that Columella calls both kinds of wine, "wine." Here is a clear ancient reference to nonalcoholic wine actually being called "wine."

"Care should also be taken so that the must, when it has been pressed out, may last well or at any rate keep until it is sold. We will then next set forth how this ought to be brought about and by what preservatives the process should be aided. Some people put must in leaden vessels and by boiling reduce it by a quarter, others by a third. There is no doubt that anyone who boiled it down to one-half would be likely to make a better thick form of must and therefore more profitable for use, so much so that it can actually be used, instead of must boiled down to one-third, to preserve the must produced from old vineyards." Continuing, "We regard as the best wine any kind which can keep without any preservative, nor should anything at all be mixed with it by which its natural savour would be obscured; for that wine is most excellent which has given pleasure by its own natural quality." [143]

The previous statement is notable. First, it tells how new wine (must) is preserved by boiling it down. Columella seems to prefer that it be boiled down to ½ its consistency. Continuing his thought about must being boiled down to ½, he then speaks of the best wine (must) as being that which needs no preservatives added to it. This most probably is because of its thick consistency that would ward off bacteria and fermentation. While some have argued that this "best wine" statement is referring to alcoholic wine, it would seem it is actually referring to thick wine or must that needs no preservatives and thus retains its natural flavors. Remember this is under the same heading speaking about boiling must. In the Loeb Classical Library edition the margin heading reads, "How to preserve and strengthen wine." [144] Also, after the "best wine" sentence, the very next line says, "But when the must..." [145] continuing this discussion of must. Apparently Columella believed the very best wine was boiled down, unfermented, with no additives!

Several other references from Columella show the plain evidence of non-alcoholic wine. "Since some people - and indeed almost all the Greeks - preserve must with salt or sea-water..." [146] "Therefore a careful landed proprietor, when he has secured an estate, will at the first vintage immediately make trial of three or four different kinds of preservatives in as many different jars, so that he may discover how much salt water the wine which he has made can stand without spoiling the taste." [147] "There is another method of preserving must by means of liquid pitch." [148]

In reference to boiling, here is what Columella has to say, "The bailiff should have in readiness logs which he may use for boiling down the must to a third or half its original volume. Also salt and spices, which he has been accustomed to use in the preservation of wine, ought to be stored up in good time beforehand." [149] "The more the must is boiled down, - provided it be not burnt - the better and the thicker it becomes." [150] "The cauldron-room, in which boiled wine is made, should be neither narrow nor dark, so that the attendant who is boiling down the must may move around without inconvenience." [151] "As to the part devoted to the storage of produce, it is divided into rooms for oil, for presses, for wine, for the boiling down of must, lofts for hay and chaff, storerooms, and granaries," [152] Notice Columella refers to boiled wine, then boiled must; he appears to use must and wine synonymously. Boiling was a normal part of the ancient wine making process. The boiling process rids wine of alcohol.

Pliny was a Roman writer who lived during New Testament times. He died in an attempt to rescue people stranded as a result of the eruption of the volcano, Mt. Vesuvius (AD 79). He wrote a virtual encyclopedia of the first century AD. Some of that encyclopedia deals with vineyards, grapes, and wine. Here is a sample of his recordings.

"Draughts of wine to restore the strength." [153] It is clear that unfermented wine has more capability to restore strength than intoxicating wine. In another place he writes, "One of the black grapes has been named 'the good for-nothing,' though it might more properly be styled 'the sober,' as the wine it produces is admirable, particularly when old, but though strong it has no ill effects: in fact this is the only vintage that does not cause intoxication."[154] The wine mentioned here is called "strong wine" causing "no ill effects" and even after aging "not cause[ing] intoxication." Some will point out that Pliny says this is the only kind that does not cause intoxication. Other authorities, and evidence, would disagree. This does not negate the fact, however, that Pliny points out an aged, un-intoxicating wine.

Pliny mentions two kinds of grapes that can be stored in a jar and, "kept in new wine boiled down and must, and properly in after-wine." [155] Pliny mentions wine called "life" that had a large amount of sea-water mixed in. He relates how some wine is called "sea-flavoured wine" and "sea-treated." Some wines were saltier than others. [156] These would have been mostly, or altogether non-alcoholic wines. Pliny mentions raisin wine, a kind of must, sapa (made by boiling wine down to a third of its quantity), defrutum (must boiled down to one-half), and then says, "all these wines." [157] So Pliny here, in contradiction to a previous statement, refers to raisin wine, must, sapa, defrutum, and boiled down wine, all as "wine."

Ancient writers sometimes indicate only intoxicating wine is real wine, but later forget their strict designation and go back to calling non-alcoholic juice "wine." This is evidenced in the next two quotes, "The liquors made from grape-skins soaked in water, called by the Greeks seconds and by Cato and ourselves after-wine, cannot rightly be styled wines, but nevertheless are counted among the wines of the working classes. Cato called a type of this, 'wine-lees.'" [158] This type wine was boiled and probably unfermented. It may not have enough sugar content to ferment in the first place. Notice he does not want to call it wine, but recognizes people count it among the wines. Pliny calls "wine" that which was made from fruit such as carob, pears, apples, pomegranates, cornels, medlars, service berries, mulberries, fir-cones." Finally he refers to "A wine is also made of only water and honey." [159] This shows how the word wine was used in such a generic and general way.

Pliny speaks of a sweet honey wine "is made from must, in the proportion of thirty pints of must of a dry quality to six pints of honey and a cup of salt, this mixture being brought to a boil; this produces a dry-flavoured liquor." [160] If the must had any alcohol content, it is excluded when this mixture is boiled. The salt preserves it from converting to alcohol. As previously mentioned, in lactic fermentation the wine ferments, but not into alcohol. This would have been a non-alcohol or very low alcohol drink.

Pliny then says, "myrrh-wine was counted not only among wines but also among sirops." [161] So, according to Pliny, myrrh wine was counted as wine, and also counted as syrup (sirops). A wine syrup, boiled down must, would not be alcoholic. "Boil tender sprigs of myrtle with the leaves on in salted must, and after pounding them boil down one pound of the mixture in 2 ¼ gallons of must until only 1 ½ gallons are left. The beverage made by the same process from the berries of the wild myrtle is called myrtle wine; this stains the hands." [162] Three factors demonstrate this is non-alcoholic wine. First, it is boiled. Second, it is boiled to a thick consistency that would retard alcoholic fermentation. Third, the salt content retards alcohol fermentation and promotes lactic fermentation.

Often Pliny speaks of the boiling of various wines. This boiling would not only preserve it, but as previously stated would increase the sugar content, thus prohibiting the production of alcohol. Notice, "Wormwood wine is made by boiling down a pound of Pontic wormwood in five gallons of must to one-third of its amount."[163] "'Sapa.' Must, or new wine, boiled down to one half, according to Pliny; and one third, according to Varro." [164] In his work *Natural History*, he mentions "a mixture of boiled wine." [165]

Some wines, while sweet and non-alcoholic, still worked on the digestive system. "Egypt gives the name of 'wine of Thasos' to an extremely sweet native vintage which causes diarrhea." [166] "As to the Surrentine wines, they have no such effect upon the stomach, nor are they at all trying to the head; they have the property also of arresting defluxions of the stomach and intestines." [167] "Sweet wine, again, is less inebriating, but stays longer on the stomach, while rough wine is more easy of digestion." [168] Note these wines are not "trying to the head" and "less inebriating."

Interestingly, we find a reference to frozen wine from Pliny and another writer. "It has been seen occasionally, that the vessels have burst in a frost, leaving the wine standing in frozen blocks - almost a miracle, since it is not the nature of wine to freeze: usually it is only numbed by cold." [169] Over three centuries earlier Zenophon wrote, "Here fell a great snow, and the cold was so severe that the water the servants brought in for supper, and the wine in the vessels, were frozen." [170] While we can't absolutely prove this is unfermented wine, the higher the alcohol content in wine, the lower the temperature has to be for it to freeze. Alcohol freezes at a lower temperature than water. This would at least lend evidence to the possibility this particular wine had no alcohol, or low levels of alcohol.

Marcus Terentius Varro lived from 116-27 BC. So he died about 22 years before Christ. Varro served as a lieutenant under Pompey and fought Mediterranean pirates. He was commissioned by Caesar to oversee the collection of a great public library. He spent the last years of his life devoted to study and writing. Varro began writing *On Agriculture* at 80 years of age.

Speaking of the vineyards and the grapes produced by them, Varro writes, "The vines can bear a large quantity of wine of good quality... while the wine is forming and ripening it does not need water, as it does in the cup, but sun." [171] Vines naturally bear nonalcoholic wine, you have to work at it to produce drinkable alcoholic wine. Varro calls "wine" that which was forming and ripening on the vine. He further indicates that "When the grapes have been trodden, the stalks and skins should be placed under the press, so that whatever must [new wine, unfermented wine] remains in them may be pressed out into the same vat." [172]

Plutarch (c. AD 46-120) was a Roman writer, historian, and biographer. He lived during New Testament times. He states, "Why does not new wine inebriate as soon as other?" [173] Plutarch then continues with a confused discussion on why new wine does not make you drunk. We would term things differently today, and we now know that alcohol is the reason for inebriation. But this discussion reveals the fact of different kinds of wine in ancient times, alcoholic and nonalcoholic. Plutarch further says, "Wine should not be heady till it hath lost its sweetness." [174] This is also an additional statement clarifying sweet wine. Wine is not heady until it has lost it sweetness through alcoholic fermentation. In this sentence he calls it wine *before* and *after* it has lost its sweetness.

Plutarch discusses straining or filtering wine, "Those that strain wine geld and emasculate it. So they take all the strength from the wine, leaving the palatableness still: as we use to deal with those with whose constitution cold water does not agree, to boil it for them. For they certainly take off all the strength from the wine, by straining of it...But for your part, you would glut yourself with night wine, which raises melancholy vapors; and upon this account you cry out against purgation, which, by carrying off whatever might cause melancholy or load men's stomachs, and make them drunk or sick, makes it mild and pleasant to those that drink it, such as heroes (as Homer tells us) were formerly wont to drink…But what hurt, I pray, have I done to the wine, by taking from it a turbulent and noisome quality, and giving it a better taste, though a paler color? But you will say that wine not strained hath a great deal more strength. Why so, my friend? One that is frantic and distracted has more strength than a man in his wits…purging of wine takes from it all the strength that inflames and enrages the mind, and gives it instead thereof a mild and wholesome temper."[175] This practice is still utilized by some modern companies today for both wine and grape juice. Examples are Draper Valley Vineyard (filters wine or grape juice before it has fermented) and Ariel (makes fermented wine, then filters out the alcohol).

Note again this quote from Andrew Dalby on Plutarch: "The power of alcoholic drinks to intoxicate was familiar enough, but the presence in them of alcohol and the nature of alcoholic fermentation were not understood; hence Plutarch's puzzled discussion of why gleukos, must or fresh grape juice, is not intoxicating." [176]

Galen (c. AD 190), a physician and philosopher, speaking of the straining quality of the kidney, says, "just as the whole of the wine is thrown into the filters." [177] Straining wine was apparently a common practice.

Virgil (70 BC - 19), previously mentioned, was the son of a farmer and a classical Roman poet. Sampling his work we find, "I will pour from goblets, fresh-strained sweet grape-juice, equal to the choice Arvisian wines of Chios' isle." [178] Fresh strained sweet wine or grape juice would be nonalcoholic. This is equaled to choice Chian wine. This also shows fresh unfermented wine was common.

Theophrastus (c. 371-287 BC) was born on the Greek island of Lesbos and studied under Plato and Aristotle. He became Aristotle's successor. In his *Inquiry Into Plants*, Book IV and elsewhere, Theophrastus often distinguishes between sweet wine and dry wine. That is a basic difference, of course, between sweet unfermented wine, must, or grape juice; and dry, intoxicating wine. He commented on various aspects of wine.

"Wine of Erythrae, which has a taste of brine, and is subtle."[179] This alludes to preserving wine with salt or lactic fermentation. "Nevertheless, to speak generally and broadly, sweet flavours and those of that order are more nutritive than the rest and more natural...The sweet flavour is agreeable to all." [180] Theophrastus mentions wine "strained through a cloth." [181] He also mentions "Thus the acid is believed to belong more to the fluid, and so the dry-wine; whereas the pungent is believed to belong more to the dry, and so the sweet (at all events fluids get sweeter as they thicken)." [182] Finally we find, "And wine that has been filled with foreign fluid from the very start of harvesting, owing to prevailing rain, and wine that in some other way has acquired a component of water, all such wines readily turn acid on the slightest occasion." [183]

If the previous examples have still not convinced the reader, the following pages are filled with references to ancient writers who used the word to refer to a non-alcoholic beverage.

Vines Producing Sweet Wine

"He planted it all over with fruitful trees and vines producing sweet wine." [184] Notice the vines produce the fruit of the vine; the natural product is sweet, unfermented wine. Press a bunch of good, ripe, fresh grapes into a cup and you will be surprised at how sweet it tastes. That sweet wine could have been preserved in that unfermented state, fermented through nonalcoholic lactic fermentation, or allowed to ferment into an alcoholic drink. But this is another ancient reference to fresh, sweet, nonalcoholic wine.

Plump Sheep and Sweet Wine

"It's true they eat plump sheep and drink the best sweet wines." [185] One of many ancient references to sweet wine. Many more such examples could be given. Yes, alcoholic wine could be made sweet (you can make alcohol taste most any way you want), but in ancient times the normal meaning of sweet wine was that which had not gone through alcoholic fermentation, thereby retaining the sugar or sweetness.

Sweet Wine & Must in Talmud

"In the chapter of the **Talmud** on 'Offerings,' sweet wine is mentioned. In the chapter on 'Vows' it is stated: 'If any one has vowed that he will abstain from wine, then there is permitted to him boiled must in which is the flavor of wine,...also cider of apples.'" [186] Note the distinction between intoxicating and un-intoxicating drinks. Also, "cider of apples" would be an un-intoxicating example of shekar.

Fruit Containing Wine

"Fruit containing wine must be given restrictedly." [187] Aretaeus was a Greek physician who lived in the first century AD.

Sweet, Mild Wine

"After Thurii comes Lagaria, a stronghold, bounded by Epeius and the Phocaeans; thence comes the Lagaritan wine, which is sweet, mild, and extremely well thought of among physicians." [188]

Fresh Squeezed Wine

Nicander, a second century BC Greek poet, states, "And Oineus squeezed it out into hollow cups and called it wine [oinos]." [189] The context makes it clear it is speaking of squeezing grapes and calling it wine. Five centuries before the birth of Christ in Bethlehem, Anacreon, a Greek poet made another clear reference to fresh grape juice being called wine, "Squeeze the grape, let out the wine [oinos]." [190]

Ancient Hittite Tablet

"[Behold] the raisin. Just as it holds its wine in [its] heart..." [191] A raisin holds unfermented wine in its heart. Raisin wine was a common way to produce unfermented wine at any time of year. McGovern also states Hittite texts were replete with references to "sweet wine."

Flavius Josephus (c. AD 37-100) was a first century Jewish historian. He lived during the time of the writing of the New Testament. His writings have greatly increased understanding of Bible times. Speaking of the Genesis 40:11 account, Josephus says, "He therefore said, that in his sleep he saw three clusters of grapes hanging upon three branches of a vine, large already, and ripe for gathering; and that he squeezed them into a cup which the king held in his hand; and when he had strained the wine, he gave it to the king to drink, and that he received it from him with a pleasant countenance." [192] While the Genesis 40:11 does not specifically call it wine, the first century AD Jewish historian Flavius Josephus does specifically call these fresh squeezed grape clusters "wine."

Josephus continues by saying Joseph "let him know that God bestows the fruit of the vine upon men for good; which wine is poured out to him, and is the pledge of fidelity and mutual confidence among men; and puts an end to their quarrels, takes away passion and grief out of the minds of them that use it, and makes them cheerful." [193] This quote immediately follows Josephus's previous quote about squeezing grape clusters into Pharaoh's cup and calling it wine. Josephus then quotes Joseph as saying that this fresh squeezed unfermented wine is given by God for good, puts an end to their quarrels, takes away passion and grief, and makes them cheerful. This is the opposite of intoxicating wine. Some may object and say Josephus is putting words in the mouth of Joseph. If that be true, it still presents the opinion of an influential Jewish historian of the first century AD. And his opinion is that fresh nonalcoholic grape juice is to be considered "wine."

Plato (c. 428 - 347 BC) was founder of the Academy in Athens and a philosopher and mathematician. He was taught by Socrates, Plato in turn taught Aristotle. Plato helped lay the foundations of Western philosophy and science. Here is his contribution, "I am also sending twelve jars of sweet wine for the children and two of honey." [194] Why send sweet wine for the children? Because it is not intoxicating. It is perfectly safe for children to drink. On the other hand, ancient writers warned against giving wine (alcoholic wine) to children.

Considering the abundant evidence above, it is untenable to maintain there was only one kind of wine in ancient times. Ancient people obviously had alcoholic wine, though they had no specific word for alcohol. They also obviously had sweet, nonalcoholic drinks they also called wine. When reading the word wine in the Bible or ancient literature, none should jump to the conclusion it was always and only strong intoxicating wine.

Modern Day Acknowledgements of Ancient Nonalcoholic Wine

A modern dictionary defines *Must* as, "n. [L. mustum, new wine, <mustus, new, fresh] unfermented juice pressed from grapes or other fruits." [195]

To those who deny must is wine, this definition points back to the idea of new wine. Must is wine, just a certain kind of wine; unfermented wine. Some have said must can be intoxicating. Many authorities would disagree. For example, an antiquities dictionary states, "Intoxication is one of the marks of the Dionysiac festival, and must cannot intoxicate, it needs fermentation." [196]

Robert P. Teachout, in his 1979 doctoral dissertation explains, "One kind of 'wine' (Greek oinos, Latin vinum) which was explicitly valued and drunk in the Graco-Roman world was unfermented grape juice. (Whereas this was, of course, also true of the early Egyptians and the

Mesopotamian peoples, the evidence for it is not as readily available or as extensive.) Not one but several different means were used to preserve the juice long after the harvest." [197]

Andrew Dalby is a historian, linguist, and author of several books on the classical world. His *Food in the Ancient World From A to Z,* is a secular, rather than a religious book. While he would disagree with much of my book, he plainly acknowledges nonalcoholic wine in the ancient world. Here are his acknowledgements. "It [Theran wine] was very sweet, quite 'thick' and black in colour, according to Galen." [198] Theran wine was sweet and quite thick; both indicators of ancient unfermented wine. "Abate wine is said by Galen to be at once austere and sweet, 'thick' and black" and "The term 'Cretan Wine' meant a very sweet wine, a passum or protropos, as is evident from Dioscorides's and Galen's uses of the word." [199] Tarentine wine is taken as "the pattern for wines of deep southern Italy: they are all 'simple,' not intoxicating, not forceful, pleasant, easy on the stomach. Horace, Odes 2.6.19-20; Pliny NH 14.69; Statius, Silvae 2.2.111; Martial 13.125; Juvenal 6.297; Athenaeus E 27c." [200] Interesting note is how many ancient sources Dalby quotes affirming this position. Also note Dalby's last quote that Tarentine wine was "not intoxicating" and was "the pattern for wines of deep southern Italy."

Andrew Dalby, in a quote previously given, points out that Plutarch knew, but did not understand why gleukos (usually translated "new wine" or "sweet wine") was not intoxicating. Gleukos is the Greek word used in Acts 2:13. We get our word glucose from this Greek word for new, un-intoxicating wine. The English word glucose does not refer back to alcoholic wine, but to the sweet, nutritious nature of the Greek wine gleukos. In the ancient world a wine could be strong, concentrated, yet still be non-intoxicating. Dalby continues, "Greek *amethystos* 'unintoxicated,' equated by Columella with Latin *inerticula nigra*. The variety was supposed to make 'strong but non-intoxicating' wine. Columella DA 3.2.24; Pliny NH 14.31" [201]

George Whitefield Samson was a graduate of Brown University and Newton Theological Seminary. He was a highly educated, respected Baptist pastor and president of George Washington University and later at Rutgers. He authored several books. He writes, "The Syriac term for the Hebrew 'yayin' is 'chamro,' corresponding to the Hebrew 'chemar,' the Chaldaic 'chamra,' and the modern Arabic 'chamer.' The Hebrew 'tirosh' is also usually rendered 'chamro' in the Syriac; 'chamro,' like the Greek 'oinos,' and the Latin 'vinum,' being the generic term." [202] Dr. Samson's point is that these words for wine were used generically and interchangeably of all kinds of wine.

Three Kinds of Wine

Lyman Abbott contends, "There is no doubt that there were three principal kinds of wine known to the ancients. First, there was fermented wine. It contained what is the only objectionable element in modern wines, a percentage of alcohol. It was the least common, and the percentage of alcohol was small...Second were the new wines. These, like our new cider, were wholly without alcohol, and were not intoxicating. They were easily preserved in this condition for several months. Third were wines in which, by boiling or by drugs, the process of fermentation was prevented and alcohol excluded." [203] Notice Dr. Abbott says there is no doubt nonalcoholic wines existed and were common. Also significant, he says alcoholic wine was least common in ancient times.

Various Ancient Cultures Word for Wine

Dr. Teachout writes, "It is also interesting that the Akkadian word for 'wine,' though not related to yayin, was used in a similar manner: both for fermented wine and for 'must' (grape juice)." [204] Jean Bottero, "Vine, grape, and wine also had one and the same name in Akkadian: karanu, an old term common to all Semitic languages to evoke 'vine' or 'vineyard.'" [205] Several sources affirm the Hittite (or Sumerian) word for wine is Gestin. Gestin had a wide generic use, somewhat like the Hebrew word yayin, the Greek oinos, the Latin vinum, the Akkadian karanu. Gestin was found to not only mean wine, but could also mean vine, vineyard, grape, or raisin. [206] Patrick McGovern adds, "One of the Egyptian hieroglyphs for 'grape, vineyard, or wine' is very similar to the Linear A and B signs of LM and Mycenaean times. The vocabulary of a technology such as winemaking - the actual script in this instance - very often follows along with its transference from one culture to another." [207] Notice that like other ancient languages, the Egyptian hieroglyph for wine, was generic. So general that it could be used for grape, vineyard, or wine. Very likely this included unfermented wine. Notice also McGovern says this vocabulary often transferred from one culture to another. Finally we see "Residue [from about 3000 BC in Egypt] in vessels found in tombs has been assumed to be either wine or grape juice." Wines in Egypt included, "a range from sweet to very dry." Athenaeus said the Mareotic wine "was remarkable for its sweetness." [208]

The evidence clearly shows the ancients often used the word "wine" to designate the unfermented nature of the beverage.

CHAPTER FOUR:
Biblical Words for Wine

The ancients had nonalcoholic wine available throughout the year. They knew how to preserve nonalcoholic wine. They referred to both alcoholic and nonalcoholic beverages as wine. These facts have already been established. It is not my intention to go into great detail on the Hebrew and Greek words for wine. The brief discussion following will only touch on the biblical words for wine. It will give evidence from leading authorities on the basic meaning of these words.

The Old Testament portion of the Bible was originally written in Hebrew. Two of the most common Hebrew words for wine in the Old Testament were *yayin* and *tirosh*. Despite some claims to the contrary, yayin was used generically to sometimes refer to alcoholic wine, sometimes to nonalcoholic wine. Some scholars claim tirosh always referred to unfermented wine or grape juice. Others say while it usually referred to unfermented wine, occasionally it was used to refer to alcoholic wine. Assuming the latter only lends evidence to one more word for wine being used in a generic sense.

The New Testament portion of the Bible was originally written in Greek. The two most common Greek words for wine were oinos and gleukos. Evidence has been, and will be shown, that oinos referred to both alcoholic and nonalcoholic wine. Gluekos generally referred to new, sweet, unfermented wine. That gluekos is unfermented wine is attested by Aristotle, Plato, Hippocrates, Plutarch and others. Also see later discussion on Acts 2:13 (Chapter 8).

Two quick examples will here suffice. Proverbs 3:10 is clearly speaking of nonalcoholic new wine, wine that has just been pressed. The Hebrew word is tirosh. Isaiah 16:10 is clearly speaking of nonalcoholic new wine. The Hebrew word Isaiah used was yayin. Both of these verses are translated into English by the word wine. The *Septuagint* (a.k.a. LXX) was a translation of the Hebrew Bible (our Old Testament) into Greek in about 200 BC. The Jewish scholars of the Septuagint translated these two verses (both tirosh and yayin) with the Greek word oinos. This shows that ancient Jewish scholars viewed nonalcoholic grape juice as oinos, or wine. This also demonstrates that even modern day English Bible translators view wine as sometimes fresh expressed grape juice (more on this in next chapter). Frankly, these two verses also reveal that God Himself views nonalcoholic grape juice as wine.

The ancient words for wine were used to refer to grapes still on the vine, fresh expressed grape juice, fermented wine, preserved unfermented wine, wine greatly watered down, and vinegar.

Some insist that because *today* wine always means an alcoholic drink, then that has to be what it meant in Bible days. Words, however, change meaning over time. While wine is still occasionally used in English today to refer to grape juice or nonalcoholic wine, this was even more common in ancient times. And if we want to properly understand the Bible, it is more important to know how the word was used then than how it is used today. Another consideration is that even in our modern day English language, most every word has more than one meaning.

Dr. Murphy put it well. "The usage of the time and place of the writer determines the meaning. If a word or phrase had several meanings, the context determines which it bears in 'a given' passage. The more common meaning of the writer's day is to be preferred, provided it suits the passage, -not that more common in our day." [209]

Quotes and references on the biblical words for wine.

Dr. John Kitto, D.D., F.S. A., was one of the most respected Bible scholars of the 1800s. He had lived in the Middle East and had an extensive knowledge of Bible times and customs. Charles H. Spurgeon highly praised Kitto's writing. Kitto's *Popular Cyclopedia* is still available and valuable today. He says, "No fewer than thirteen distinct Hebrew and Greek terms are rendered in our common version by the word 'wine.' Besides the pure juice of the grape, frequent mention is made in Scripture of a kind of boiled wine or syrup, the thickness of which rendered it necessary to mingle water with it previously to drinking (Proverbs 9:2,5), and also of a mixed wine, made strong and inebriating by the addition of drugs, such as myrrh, mandragora, and opiates (Proverbs 23:30; Isaiah 5:22). This custom has prevailed since the earliest ages and is still extant in the East. We are not, however, to conclude that all mixed wine was pernicious and improper. There were two very opposite purposes sought by the mixture of drinks. While the wicked sought out a drugged mixture, and was 'mighty to mingle strong drink,' Wisdom, on the contrary, mingled her wine with water or with milk (Proverbs 9:2,5) merely to dilute it and make it properly drinkable. Of the later mixture Wisdom invites the people to drink freely, but on the use of the former an emphatic woe is pronounced. In Isaiah 25:6, mention is made of 'wines on the lees.' The original signifies 'preserves' or 'jellies,' and is supposed to refer to the wine cakes which are esteemed a great delicacy in the East." [210] Notice that Dr. Kitto referred to the pure juice of the grape, and boiled wine or syrup as wine.

Wine a Generic Word

Dr. Adam Clark, a Methodist scholar and author, writes, "The Hebrew, Greek, and Latin words which are rendered 'wine,' mean simply the expressed juice of the grape." [211] In other words, Hebrew, Greek, and Latin words for wine meant expressed juice of the grape, whether alcoholic or not. Vinegar is called wine (sour oinos) in Mark 15:23. Aristotle says, "Wine and water is called 'wine.'" [212] Dr. Jim Richards speaks to this fact when he says, "Word studies in the original language point to the possibility of generic usage of the words translated 'wine.' There are scholars who say 'yayin' and 'tirosh' (Hebrew) and 'oinos' (Greek) can mean non-fermented, non-alcoholic drink (Isaiah 16:10, Joel 1:10). Tirosh in some English versions is translated "grapes" (Micah 6:15). In the same verse Yayin is translated 'wine.'" [213] Furthermore, "Cider is an English word that can mean alcoholic or non-alcoholic juice. 'Wine' as translated in most English versions could refer to fermented or non-fermented liquid. There is biblical evidence that the ancients drank non-alcoholic 'wine' or what we would call 'grape juice.' In Deuteronomy 32:14, Moses said the Israelites drank the "blood of the grape," not an alcoholic beverage. Perhaps the greatest example of grape juice use is found in Genesis 40:11. Here grapes are pressed into a cup and given to Pharaoh to drink. This 'pure blood of the grape' could not have been alcoholic." [214]

Wine like the Word Drink

Dr. Sutton is a pastor, author, and historian. He offers the following argument: "You are as familiar as I am with the fact that wine is like the word 'drink' in our day. I can say I'm going to go have a drink after work. It could be a lemonade or it could be a beer. Wine was a very common term in that day. I could take this glass right here and have a bunch of grapes in my hand and squeeze them and take the juice, and I would have wine, biblically speaking." [215]

Oinos

Dr. Stephen Reynolds was a Presbyterian scholar, held a Ph.D. from Princeton University, and worked in biblical and oriental languages. He contributed together with other faculty members of Gordon-Conwell Theological Seminary to the translation of the New International Version of the Old Testament, and wrote articles in *Baker's Dictionary of Ethics* edited by Dr. Carl F. H. Henry. Speaking of oinos he writes, "IN AS MUCH as the Greek word *oinos* is the translation in the Septuagint of the Hebrew word *yayin* when the latter clearly means the freshly pressed juice of the grapes, this is proof that *oinos* may mean unfermented grape juice. See Isaiah 16:10 "the treaders shall tread out no *yayin* (Hebrew), *oinos* (Greek) in their presses." The meaning of both *yayin* and *oinos* is obviously unfermented grape juice. In Proverbs 3:10 the freshly pressed juice of the grape is also called *oinos* in this same Greek version of the Hebrew Bible, a version which the inspired New Testament writers frequently quoted. In this passage, where the King James Version reads "thy presses shall burst out with new wine" (*new wine* translating Hebrew *tirosh*) the Septuagint simply uses the word *oinos* without any adjective *new*. What comes from the presses is not alcoholic!...This should be enough to establish the idea that wherever *oinos* appears in the New Testament, we may understand it as unfermented grape juice unless the passage clearly indicates that the inspired writer was speaking of an intoxicating drink." [216]

Likewise, Dr. Kenneth Gangel states, "Oinos...is used for both intoxicating and non-intoxicating wine." [217] Young's Concordance refers to oinos as "wine, grape juice." [218] Notice that for the Hebrew word yayin, and the Greek word oinos, Young describes them simply as what is pressed out of grapes; that could be either alcoholic or not.

Yayin

Yael Zisling explains, "The word *yayin* was used to indicate fermented or unfermented wine." [219] This is a simple quote from a pro-fermented Israeli wine site, but it recognizes that yayin was a generic word. The article continues, "Therefore, it's no surprise that wine is being exported from

Israel, today. However, the fact that some truly good wine is produced here still seems to elude many. Interestingly, part of the industry's former export success, lies with the export of non-alcoholic 'grape juice' to Moslem markets…The Talmud mentions more than a dozen types of wine." [220]

Two Jewish Encyclopedias state, "Fresh wine before fermenting was called 'yayin mi-gat' (wine of the vat; Sanh. 70a)." [221] "The newly pressed wine prior to fermentation was known as yayin mi-gat." [222] Notice these Jewish Encyclopedias use both "wine" and "yayin" to refer to unfermented wine or grape juice. Turning again to Young's Concordance we find, "Yayin - what is pressed out, grape juice." [223]

Dr, Robert Teachout's doctoral dissertation of the subject has this to say, "Yayin can refer either to fresh juice or to fermented wine…That is, not the strained interpretation of a few verses or some doubtful etymological data, but a host of strong evidence points to the fact that yayin can mean not only wine but also grape juice, and that God's purpose for the vine was the latter." [224]

Hamar

On another ancient word for wine, Dr. Teachout demonstrates that, "Hamar then is an exact synonym for yayin in that it can be designated either a fresh juice or a fermented wine." [225]

Gleukos

The Dictionary defines the Greek word from which we get our word glucose as "must, sweet new wine." [226] Young's Concordance says, "Gleukos - sweet or new wine." [227]

Dr. G. W. Samson, author of *The Divine Law as to Wines*, says gleukos, "has from such writers as Hippocrates and Aristotle been shown to be must, or preserved grape-juice…The neuter noun glukos, contrasted as to its medicinal qualities by Hippocrates, the earliest Greek medical writer, with 'sweet wine,' is wine in which the first ferment has been prevented, so that it is the Latin mustum, or unfermented grape-juice." [228]

Dr. Steven Reynolds contributes, "Gleukos refers to the sweetness of unfermented grape juice before the yeast has destroyed the natural sugar (Galen uses this word in the sense of grape juice. Vol. 6). Wine with alcoholic content can be sweetened by mixing in sugar before or after the microorganisms which produce the alcohol have died, but scholars do not accept the idea that such artificially sweetened wine was the meaning of gleukos." [229]

Tirosh

The Jewish Encyclopedia states, " 'Tirosh' includes all kinds of sweet juices and must, and does not include fermented wine (Tosef., Ned. iv. 3)." [230] Some would disagree and say there are a few exceptions and tirosh is sometimes alcoholic. If so, that just serves as another example of a word for wine referring to the juice of the grape, whether fermented or unfermented. Still, it is significant that the Jewish Encyclopedia says tirosh does not include fermented wine, and that the Jewish Septuagint translates tirosh into the Greek word oinos. This shows that ancient Jews had available unfermented beverages. It shows "must" was usually, if not always, unfermented. It is also noteworthy that tirosh is often translated with the word "wine" in modern English Bible translations.

Asis

Brown, Driver, and Briggs define an additional Hebrew word for wine, "Asis" as "freshly pressed out, sweet, not yet fermented juice of the grapes." [231]

Shekar –"Strong Drink" and Wine

Dr. Lyman Abbott (1835-1922), was a Bible scholar, Congregational minister, editor, and author. He records, "It is tolerably clear that the general words 'wine' and 'strong drink' [shekar] do not necessarily imply fermented liquors, the former signifying only a production of the vine, the latter the produce of other fruits than the grape." [232]

John W. Haley writes this, "Certain authors maintain, with some plausibility, that in all cases where strong drinks are coupled with terms of commendation, the original word properly means either *unfermented wine* or else *fruit*; and that the notices of fermented wine are restricted to passages of a condemnatory character. This position, if tenable, is one of great importance." [233] Haley then footnotes the *Temperance Bible Commentary* and other sources.

If the Common Wine was Technically Alcoholic...

For those who insist the common wine in Bible times was alcoholic, they still have a big problem. Dr. Robert Stein, who taught at Bethel College and has a Ph.D. from Princeton Seminary, considers that even if common wine was alcoholic and mixed with great amounts of water, it was hardly comparable to modern day alcoholic beverages. "To consume the amount of alcohol that is in two martinis by drinking wine containing three parts water to one part wine, one would have to drink over twenty-two glasses. In other words, it is possible to become intoxicated from wine mixed with three parts of water, but one's drinking would probably affect the bladder long before it affected the mind" [234] Dr. John A. Broadus, a Baptist theologian and author, takes this same general approach. He agrees that wine regularly used in the New Testament, "would stimulate about as much as our tea and coffee." [235]

Charles Wesley Ewing points out not all wine is a mocker. "'Wine is a mocker, strong drink is raging: and whosoever is deceived thereby is not wise' (Proverbs 20:1). Sweet unfermented wine is no mocker. It is not raging. It does not deceive the user. Fermented wine is the mocker, rager, and deceiver." [236]

Ferrar Fenton, by age 28, had acquired a working knowledge of 25 classical, Oriental, and modern languages. He states, "It should never be forgotten that when reading the Bible and the classic pagan writers of 'Wine,' we are seldom dealing with the strongly intoxicating and loaded liquids to which that name is alone attached in the English language, but usually with beverages such as above described. They were as harmless and sober as our own Teas, Coffees, and Cocoas. Had they not been so, the ancient populations would have been perpetually in a more or less state of drunkenness...These facts should never be forgotten when we read of 'Wine' there, - for it was simple fruit syrup, except where especially stated to be of the intoxicating kinds." [237]

Dr. Moses Stuart was a graduate of Yale, Congregational pastor, author, and professor at Andover Theological Seminary. He taught the famous missionary and translator, Adoniram Judson. He pronounces, "I regard it as established beyond fair contradiction, that it was a very common thing to preserve wine in an unfermented state, and that when thus preserved it was regarded as of a higher and better quality than any other." [238] Dr. Stuart includes, "My final conclusion is this, namely, that, whenever the Scriptures speak of wine as a comfort, a blessing, or a libation to God, and rank it with such articles as corn and oil, they mean, they can mean, only such wine as contained no alcohol that could have a mischievous tendency; that, whenever they denounce it, prohibit it, and connect it with drunkenness and reveling, they can mean only alcoholic or intoxicating wine." [239]

Dr. R. L. Sumner, an evangelist, author, and editor of *The Biblical Evangelist* says, "The way fermented and unfermented wine are portrayed in the Word of God ought to tell us something. Unfermented wine is used in Scripture repeatedly as a symbol, a type of God's blessing. On the other hand, when fermented wine is used symbolically it is a type of God's wrath. One pictures blessing; the other portrays cursing. It is that way in real life, too." [240]

Finally, in the words of Dr. John R. Rice, founder and editor of *Sword of the Lord*, we find, "In Bible times they did not have distilled whisky as we have it now. However, they did have several kinds of wine. But wine in the New Testament very often means simply grape juice as we have. When the juice was first squeezed out of the grapes, it was called wine, as you see from Proverbs 3:10...So grape juice is wine, in the Bible sense. Later when the grape juice ferments, it is still called wine in the Bible sense." [241]

CHAPTER FIVE:
Evidence of Non-Alcoholic Wine in the Bible and in the English Language

We have already seen that in the ancient world the word wine was used to refer to nonalcoholic as well as alcoholic wine. The only way to know for sure what kind of wine is being referred to is to look at the context. The context usually makes plain whether the wine was fermented or not. It has also been demonstrated that nonalcoholic wine could easily be preserved in the ancient world. It has been demonstrated that ancient literature often referred to nonalcoholic wine.

Now we come to the evidence of nonalcoholic wine in the Bible itself. Does such a thing exist? It should be emphasized again that if nonalcoholic wine was common in the surrounding world during the time of the Bible, then it should have been common among the biblical people as well. While Hebrew, the language of the Old Testament, was not as well known in the surrounding world, Greek, the language of the New Testament, was very well known. The Bible used the common language of the day. If it was common for the word wine to be used generically in the surrounding world, it is likely the Bible used the words for wine in the same general way.

It should be remembered that the Bible as well as the ancient world did not have a word for alcohol. So you have to search for other details to determine if the wine is alcoholic or not. For example, in Proverbs 23:29-35 Solomon gives a detailed discussion of the results of alcoholic wine and then says not to even look at it.

There is evidence in abundance in the Bible for unfermented, nonalcoholic wine. At the outset it should be said that wine was used much like the word "drink" today. "Don't Drink and Drive" is a common slogan used by police. But it speaks to alcoholic drinks, not soft drinks. Sometimes in the Bible the word wine is a little ambiguous. But usually it is easy to tell if the wine is alcoholic or not. Also, while many assume when the Bible says wine, it is always referring to alcoholic wine, that is a false assumption. It would be more correct to assume the wine in the Bible is nonalcoholic, unless proven otherwise.

Many following verses are given that call "wine" that which is clearly nonalcoholic. In addition notice how many modern day English Bible translations also call "wine," that which is clearly nonalcoholic grape juice. So sometimes in spite of ourselves, today we continue referring to wine in a generic sense.

Was there alcoholic wine in the Bible? Of course. Everyone is agreed on that. This chapter's primary purpose, however, is to present biblical evidence that much of the wine in the Bible is referring to nonalcoholic wine -- or grape juice. Scripture in this book is from the New King James Version unless otherwise noted. The reader is welcome to study these verses in other translations as well.

The Biblical Evidence

Treading Wine

"In those days I saw people in Judah treading wine presses on the Sabbath, and bringing in sheaves, and loading donkeys with wine, grapes, figs, and all kinds of burdens." -Nehemiah 13:15 "Gladness is taken away, and joy from the plentiful field; in the vineyards there will be no singing, nor will there be shouting; no treaders will tread out wine in the presses; I have made their shouting cease." -Isaiah 16:10 You don't tread alcoholic wine in the winepress, you tread new, sweet, nonalcoholic wine.

Vats, Winepresses Overflowing with Wine

"So your barns will be filled with plenty, and your vats will overflow with new wine." -Proverbs 3:10 Vats, or winepresses overflow with unfermented wine. "The threshing floors shall be full of wheat, and the vats shall overflow with new wine and oil." -Joel 2:24 This is another instance of the Bible, and English translations, calling "wine" what is a clearly nonalcoholic beverage. Vats overflow with fresh new unfermented wine.

New Wine

New wine referred primarily, if not exclusively to fresh, unfermented wine. This new wine could easily be kept alcohol free. This type wine was common in Bible times. "All the best of the oil, all the best of the new wine and the grain, their firstfruits which they offer to the LORD, I have given them to you." -Numbers 18:12 "For she did not know that I gave her grain, new wine, and oil." -Hosea 2:8 "Therefore I will return and take away My grain in its time and My new wine in its season." -Hosea 2:9

"And He will love you and bless you and multiply you; He will also bless the fruit of your womb and the fruit of your land, your grain and your new wine and your oil, the increase of your cattle and the offspring of your flock, in the land of which He swore to your fathers to give you." -Deuteronomy 7:13 "But the vine said to them, 'Should I cease my new wine, which cheers both God and men?'" -Judges 9:13 What kind of wine cheers both God and men? New, unfermented wine. Notice the *vine* says shall I cease *my* new wine. A vine does not possess fermented wine; a vine only produces and possesses unfermented wine. If you want alcoholic wine, you have to produce it after it has been taken from the vine.

But What About These Verses on New Wine?

"Harlotry, wine, and new wine enslave the heart." -Hosea 4:11 Two references to wine are made, "wine," and "new wine." Fermented, alcoholic wine can obviously enslave the heart through addiction. New wine as well as other luxuries and delicacies can also enslave the heart. Many things today, not sinful in and of themselves, can enslave the heart; TV, computers, sports, recreation, money, work.

"Awake, you drunkards, and weep; and wail, all you drinkers of wine, because of the new wine, for it has been cut off from your mouth." -Joel 1:5 New wine normally refers to unfermented wine. This could be an exception that is alcoholic. Or it could simply refer to drunkards who drink alcoholic wine, and also gorge on food and other types of nonalcoholic sweet drinks. Even drunkards also drink nonalcoholic beverages. "Others mocking said, 'They are full of new wine.'" -Acts 2:13 Some have pointed to this verse as proving new wine (Greek word, gluekos) was highly alcoholic. However, this is the same word for wine that ancient writers (Aristotle, Hippocrates…) used to refer to wine that did not intoxicate. The more obvious interpretation of this verse is those mocking the disciples were using sarcasm or irony. It would be like someone today mocking another by saying, "He is drunk on sweet tea," or, "He is full of Dr. Pepper®." (See more detailed discussion in chapter 8, *Controversial Biblical Passages Dealing with Wine*.)

Sweet Wine

Sweet wine is often mentioned in the Bible. While there could be exceptions (alcoholic wine could later be sweetened with concentrated must, etc.), the normal, natural meaning of sweet wine was unfermented wine. Alcoholic fermentation takes away the sweetness, the sugar content. So the alcoholic product has a somewhat bitter taste. Ancient sources agree that sweet wine was nonalcoholic. Aristotle said that sweet wine would not intoxicate. Hippocrates said that sweet wine affected the head less. So when you read sweet wine in the Bible or ancient literature, unless there is good reason to believe otherwise, it should be assumed that it is speaking of sweet, unfermented, nonalcoholic wine.

"Behold, the days are coming," says the LORD, "When the plowman shall overtake the reaper, and the treader of grapes him who sows seed; the mountains shall drip with sweet wine, and all the hills shall flow with it." -Amos 9:13 (Hebrew - asis) "Go your way, eat the fat, drink the sweet, and send portions to those for whom nothing is prepared; for this day is holy to our Lord. Do not sorrow, for the joy of the Lord is your strength." -Nehemiah 8:10 Un-intoxicating wine was sweet. This is a commandment from Nehemiah for the people to celebrate a day holy to the Lord by drinking wine, sweet nonalcoholic wine. The ESV translates it, "drink sweet wine." This verse and others would also lend credence to the fact that sweet, un-intoxicating wine was common in Bible times.

But What About This Verse on Sweet Wine?

"And they shall be drunk with their own blood as with sweet wine." -Isaiah 49:26 The basic meaning of the word drunk means to be filled. If you are filled with alcohol, of course, you are intoxicated. But you can also be filled with water or orange juice. Some would gorge themselves on un-intoxicating wine; just as some would gorge themselves on food. Many in ancient times craved the sweetness of un-intoxicating wine. The Septuagint translates it, "And they shall drink - as (if it were) new wine - their own blood, and shall be filled full."

Winepress

A winepress does not press (or express) fermented wine. It presses sweet, new, unfermented wine or grape juice. That juice is called wine. And the press that never pressed alcoholic wine, is called a "wine" press. This is just another example of how the word wine is used in a very general way. "Wine" and "winepress" was used generically then, and even today.

A couple of many verses referring to the winepress follow. "And your heave offering shall be reckoned to you as though it were the grain of the threshing floor and as the fullness of the winepress." -Numbers 18:27 "Joy and gladness are taken from the plentiful field and from the land of Moab; I have caused wine to fail from the winepresses; no one will tread with joyous shouting— not joyous shouting!" -Jeremiah 48:33 Notice that new unfermented juice coming from the winepresses is also called "wine." Significant also is that a modern English Bible translation clearly calls "wine" that unfermented juice flowing in the winepress.

Wine Associated With Food

Over and over the Bible refers to wine along with other food staples. Wine is not viewed in these instances as a hard drug or an adult beverage. Rather it is simply viewed as one of the essential food items in a society whose life depended not on drugs, but on basic food and drink. The reader should be reminded of a previous chapter that explains numerous ways they had of preserving drug free wine. Unfermented wine has much more food value than alcoholic wine. Alcoholic wine is primarily a drug, rather than food; nonalcoholic wine is primarily a food.

"Therefore may God give you of the dew of heaven, of the fatness of the earth, and plenty of grain and wine." -Genesis 27:28 "Moreover those who were near to them, from as far away as Issachar and Zebulun and Naphtali, were bringing food on donkeys and camels, on mules and oxen—provisions of flour and cakes of figs and cakes of raisins, wine and oil and oxen and sheep abundantly, for there was joy in Israel." -1 Chronicles 12:40 Some say the only kind of joy from wine is that which comes from inebriation. But notice this verse refers to flour, figs, raisins, wine, oil, oxen, and sheep bringing joy. If you had nothing to eat, these would all bring you wonderful joy as well. In an agricultural society, these would especially bring elation. Just ask any gardener or farmer about the joy of the harvest.

"Therefore they shall come and sing in the height of Zion, streaming to the goodness of the LORD— For wheat and new wine and oil, for the young of the flock and the herd; their souls shall be like a well-watered garden, and they shall sorrow no more at all." -Jeremiah 31:12 "The LORD will answer and say to His people, "Behold, I will send you grain and new wine and oil, and you will be satisfied by them; I will no longer make you a reproach among the nations." -Joel 2:19 "The threshing floors shall be full of wheat, and the vats shall overflow with new wine and oil. -Joel 2:24" The foods of wheat, wine, oil are mentioned together. New wine specifically refers to un-intoxicating wine. Here wine is treated as just another food along with wheat and oil.

"The threshing floor and the winepress shall not feed them, and the new wine shall fail in her."-Hosea 9:2 Alcoholic wine is not a nourishing food; nonalcoholic sweet wine would be a very nourishing, healthy food. To lend further evidence that this is un-intoxicating wine, notice that it mentions the winepress (nonalcoholic wine flows from the winepress), and it refers to new wine.

Wine Offered to God

"And you shall bring as the drink offering half a hin of wine as an offering made by fire, a sweet aroma to the LORD." -Numbers 15:10 (also Leviticus 23:13) While the Bible may not spell it out, the implication is that the wine offered to God is sweet, unfermented wine. Following are a few reasons for this view:

1. Nonalcoholic wine was the most common, most easily preserved wine of the day.

2. Priests are specifically commanded not to drink wine when they are serving in the tabernacle. Why would God forbid to priests the very thing He asks His people to give to Him as an act of worship? Especially when a portion of the offering was given to priests for their service.

3. They were to offer wine to God as a sweet aroma. While this may not prove the point, the sweet aroma implies the offering is sweet, unfermented wine. Would a dangerous drug be a sweet aroma to God?

4. The blood sacrifices and grains offered to God were to be young, fresh, and the best. Why should the offering of wine be an offering of old, fermented, decomposed wine that would make people lose their good judgment? God never requested old, rotten grain and decomposed meat as a sacred offering. Why would He do so for wine?

5. God specifically requests the best of the new wine (Numbers 18:12; Nehemiah 13:12). This would be a clear reference to new, unfermented wine. God never specifically requests an offering of intoxicating wine.

6. Leaven was often viewed as a symbol of sin (Matthew 16:12; Luke 12:1; 1 Corinthians 5:6-8). Yeast is leaven. It is the substance that causes alcoholic fermentation. Why would God request as an offering wine with leaven?

7. Why would God ask as an offering, that which He forbade the people to partake of (Proverbs 20:1; 23:29-35)? Why would God request a hard drug as a sacred offering? Would God be impressed today with an offering of marijuana or cocaine?

Somewhat related to offering of wine to the one true God, at one time the ancient Greeks would not offer intoxicating wine to their gods. "Among the Greeks, those who sacrifice to the sun, make their libations of honey, as they never bring wine to the altars of the gods; they affirming, that it is fitting that the god who keeps the whole universe in order, regulating everything, and always going round and superintending the whole, should in no manner be connected with drunkenness." [242]

Firstfruits of New Wine

"The firstfruits of your grain and your new wine and your oil, and the first of the fleece of your sheep, you shall give him." -Deuteronomy 18:4 They were to give the priests first fruits of new wine. That new wine would have been prepared and preserved most likely by the givers, or possibly by the priests. First fruits were just harvested; they had to be immediately preserved to prevent them from spoiling. "Then all Judah brought

the tithe of the grain and the new wine and the oil to the storehouse."
-Nehemiah 13:12 New wine was usually sweet, nonalcoholic wine.

Many Kinds of Wine

"Now that which was prepared daily was one ox and six choice sheep.
Also fowl were prepared for me, and once every ten days an abundance
of all kinds of wine." -Nehemiah 5:18

Many social drinkers today insist there was only one kind of wine
in Bible times, hard alcoholic wine. Ancient people were much more
knowledgeable and industrious than we give them credit. They knew
well how to make numerous variations of drinks. They had numerous
varieties of nonalcoholic wine. They had numerous kinds of grapes
and other fruit. They knew how to preserve grain and meat; they also
knew how to prepare and preserve fresh nonalcoholic wine. Contrary
to the one-wine theory, Nehemiah had an "abundance of all kinds of
wine." This was likely many kinds of sweet wine (and possibly other
kinds of sweet fruit juice), since that is what he encouraged his people
to drink (8:10). It may have also included vinegar and other products
of the vine.

Fresh Pressed Wine

"Then Pharaoh's cup was in my hand; and I took the grapes and
pressed them into Pharaoh's cup, and placed the cup in Pharaoh's hand."
-Genesis 40:11 While this verse does not use the word wine, it is a clear,
unmistakable example of fresh, new, nonalcoholic wine. This was a common method of making wine on the spot. As explained elsewhere, this
could be done at any time of the year. The first century AD Jewish historian Josephus writes of Pharaoh's fresh pressed, nonalcoholic drink,
and he does call it "wine" (see previous chapter).

Strong Drink, Similar Drink, Bitter Drink (Shekar)

"They shall not drink wine with a song; strong drink [shekar] is bitter to those who drink it." -Isaiah 24:9 Many say shekar (Hebrew word, spelled variously) always referred to an alcoholic strong drink. Shekar, however, was made like wine, just of other fruit besides grapes. If it were prepared in a nonalcoholic way, it would be a sweet drink; in an alcoholic way it would be bitter.

This verse says your strong drink (shekar) has become bitter. The implication is, this shekar was normally sweet, but because of God's judgment, it would now be bitter. (See fuller discussion of shekar in chapter, *Controversial Biblical Passages Dealing with Wine*). Shekar, like wine, could be either an alcoholic or a nonalcoholic drink. It is also interesting that wine and shekar are often listed together as parallels. They can both be good, they can both be bad. This seems to be an instance (like Deuteronomy 14:26) of un-intoxicating shekar.

Vineyard of Wine

"A vineyard of red wine!" -Isaiah 27:2-3 The vineyard is said to contain wine. And that wine is certainly not fermented. There is a slight difference in ancient biblical manuscripts here. If the Hebrew word "khemed" is used, it means something like a pleasant vineyard. If the Hebrew word "khemer" is used, it means "wine." Notice the difference of only one letter. While modern English translations differ, the NASB, ASV, KJV, NKJV, and Darby Translation all translate this verse with the word wine.

"As the new wine is found in the cluster, and one says, 'Do not destroy it, for a blessing is in it.'" -Isaiah 65:8 Wine is in the cluster! This is an unmistakable reference to unfermented wine being called "wine." The juice still contained in the un-pressed grape is called "wine." Also, note that a modern English translation (NKJV) called "wine" what is clearly un-intoxicating wine. Other English translations that use "wine" for this clearly nonalcoholic wine: KJV, NASB, ESV, ASV, YLT, DT (Darby), HCSB. These English translations are correct. The best translation for the biblical words for wine is simply "wine." Then let the Bible student determine whether it is nonalcoholic wine or alcoholic wine by the context. But the fact of both kinds of wine should be noted in Bibles, Bible Dictionaries, and Commentaries.

Gathering Wine

"But you, gather wine and summer fruit and oil, put them in your vessels...and gathered wine and summer fruit in abundance." -Jeremiah 40:10, 12 The Hebrew word used here is yayin. The Septuagint translates it into the Greek, oinos. You do not gather fermented, but unfermented wine in the field. This terminology is similar to other ancient literature such as, "At the time of festivals, he went about, and took wine (oinos) from the fields." [243]

Infants Crying For Wine

"Because the children and the infants faint in the streets of the city. They say to their mothers, "Where is grain and wine?"" -Lamentations 2:11-12 Children and even infants are crying for wine. This is clearly a case of the common, un-intoxicating wine of the day. Surely the Jewish parents were not giving their babies a hard drug to drink. Good parents do, however, give nutritious fruit juice to infants. Plato spoke of sending jars of sweet wine to children. [244] This would have been the un-intoxicating, safe kind. On the other hand, ancients recognized the harm of giving intoxicating wine to children. "Congruity and probability are alike shocked by supposing that little children would cry out to their mothers for intoxicating drink." [245]

Daniel

"But Daniel purposed in his heart that he would not defile himself with the portion of the king's delicacies, nor with the wine which he drank." -Daniel 1:8 "I ate no pleasant food, no meat or wine came into my mouth, nor did I anoint myself at all, till three whole weeks were fulfilled." -Daniel 10:3 Daniel refused wine earlier, but later he apparently drinks wine on a regular basis. Is he a hypocrite? Did he turn his back on his previous commitment to God? No, Daniel was still a godly man. Daniel was still consistent to his convictions. The answer is simple. Daniel refused to drink the intoxicating wine of a pagan king, but he had no problem drinking the nourishing un-intoxicating wine of the day. Apparently the wine which the king drank, later had a part in bringing down Nebuchadnezzar's grandson, King Belshazzar.

Wineskin in Smoke

"For I have become like a wineskin in smoke, yet I do not forget Your statutes." -Psalm 119:83

This refers to a common method of making un-intoxicating wine. **Aristotle** (c. 350 BC) said the wine of Arcadia was "so dried up in its skins by the smoke that you scrape it to drink." [246] Wine reduced to this thick a consistency was nonalcoholic. In a chapter on wine in the Holy Land, Patrick E. McGovern reveals, "Other late Iron Age jugs and jars are inscribed with terms such as 'smoked wine' and 'very dark wine.' If we project later practice back into the past, we can understand 'smoked wine' to mean either a 'cooked' wine, in which the wine was concentrated down to a sweet syrup (Latin *sapa* or *defrutum*), or a wine that was prematurely aged by being stored in a room above a fire where the smoke could permeate the wine." [247] This method was not only used to dry and flavor wine, but also raisins. "Smoked raisins were a Roman favourite. These strongly flavoured dried grapes came from the Alban and Raetic vineyards (the latter being the ones that Augustus liked); also from North Africa, and these were the ones preferred by Tiberius. Horace, Satires 2.4.72; Pliny NH 14.16" [248]

The Simple Invited to Drink Wine

"Whoever is simple, let him turn in here!" As for him who lacks understanding, she says to him, "Come, eat of my bread and drink of the wine I have mixed." -Proverbs 9:4-5 In Proverbs 9 wisdom (personified as a woman) mixes her wine. Wisdom then sends out messengers inviting the "simple" and those who "lack understanding" to come eat her bread and drink her wine.

Why in the world would a wise person invite a man who is simple and lacks understanding to drink alcoholic wine, a hard drug? That would take away what little understanding they possessed. Remember this Scripture is divinely, infallibly inspired. Remember this same book of Proverbs condemns alcoholic wine (20:1; 23:29-35). This is an obvious reference to new wine, sweet wine, the wine that does not take away the little good judgment possessed by a simple man who lacks understanding.

Blood of Grapes

"And you drank wine, the blood of the grapes." -Deuteronomy 32:14 The reference, "blood of grapes," alludes to fresh, pure, safe wine. Cut a person and he bleeds; cut or press a grape and it bleeds unfermented wine. Jim Richards contends, "In Deuteronomy 32:14, Moses said the Israelites drank the "blood of the grape," not an alcoholic beverage." [249]

Loving Wine and Oil

"He who loves pleasure will be a poor man; he who loves wine and oil will not be rich." -Proverbs 21:17 Note that it is just as possible to love oil as wine. This is true whether wine is intoxicating or not.

Salt & Wine

"And whatever they need—young bulls, rams, and lambs for the burnt offerings of the God of heaven, wheat, salt, wine, and oil, according to the request of the priests who are in Jerusalem—let it be given them day by day without fail." -Ezra 6:9 Salt was often used to preserve wine. Through lactic fermentation the wine indeed fermented, but it was a nonalcoholic fermentation. The salt preserved nonalcoholic wine. While this verse does not specifically say the salt was used to preserve wine, it is interesting that the two are found together. The Roman writer Cato gives recipes for preserving wine with salt.

Dripping With Wine

"The mountains shall drip with new wine, the hills shall flow with milk." -Joel 3:18 Vineyards were often located on the sides of mountains or hills. Someone may say this is simply a poetic use of the word wine. Even so, this demonstrates the wide use of the word wine.

The Drinking of Joseph and Benjamin

"Then he took servings to them from before him, but Benjamin's serving was five times as much as any of theirs. So they drank and were merry with him." -Genesis 43:34 Was Joseph trying to get his little brother drunk? Of course not. You can drink and be merry without drugs being involved. You can drink to your fill and not be harmed. During at least some of ancient Egyptian history, kings and officials were not allowed to drink intoxicating beverages. "In the Cambridge Essays (1858) there is a curious paper by Mr. C. W. Goodwin, the Egyptologist, who furnished translations of some writings of a supposed very high antiquity. Several are believed to be as old as the time of Moses, and in one of them, Amen-em-an, a steward of the royal house, writes to Pentaour, a poet, in the language of reproof. Among other things he says, 'If thou wieldest the rod of office (?), men run away from thee. Thou knowest that wine is an abomination. Thou hast taken an oath (pledge?) concerning strong drink, that thou wouldst not put it into thee. Hast thou forgotten thy resolution?'" [250] Thus Joseph's offering his brother wine may have been the kind of wine found in Genesis 40:11.

Wine for the Faint

"The donkeys are for the king's household to ride on, the bread and summer fruit for the young men to eat, and the wine for those who are faint in the wilderness to drink." -2 Samuel 16:2

Alcoholic wine is void of sugar and as a drug it is a depressant. Nonalcoholic wine or grape juice with its sugar and food content would be much more nourishing to the faint. While the ancients did not understand the details of alcohol and a drug being a depressant, through trial and error, they would have known what kind of wine is more nourishing. Compare this verse to the following quotes. Pliny, "Draughts of wine to restore the strength." [251] Theophrastus, "Nevertheless, to speak generally and broadly, sweet flavours and those of that order are more nutritive than the rest and more natural." [252] Hippocrates, "Drinking strong wine cures hunger." [253]

Strong wine is not necessarily strong alcoholic wine. Boiled down sweet, un-intoxicating wine is thick and strong flavored. This kind of wine would be much more nutritious. Dr. Robert Teachout explains, "Medically, it would be foolish to give an intoxicating beverage to one who is weary and faint, since wine is a depressant. Grape juice, on the other hand, would be both refreshing and invigorating to a weary traveler along with a meal. Experience, if not medical technology, could well have taught the ancients this same truth." [254]

Drink Wine Deeply

"I have drunk my wine with my milk. Eat, O friends! Drink, yes, drink deeply." -Song of Solomon 5:1 Would it be wise to tell your friends to drink deeply of alcoholic wine? Would the Bible give such advice? Of course not; the Bible condemns drunkenness. This is another of many examples where the wine is clearly the common un-intoxicating wine of the day. Mixing wine or grape molasses with milk makes a rich refreshing safe drink.

Young Women Thrive on New Wine

"Grain shall make the young men thrive, and new wine the young women." -Zechariah 9:17 Just as young men thrive on grain, young women thrive on new wine. Wine is presented here as a normal food, just like grain. It is not presented as a recreational drug. Frankly, young women do not thrive on alcoholic wine. Recent studies have even shown that women who drink moderately significantly increase their risk of cancer. A host of other problems accompany women who drink. But young women do thrive on new wine or grape juice; it's nutritious and health benefits are numerous.

New Wine in Old Wineskins

"Nor do they put new wine into old wineskins, or else the wineskins break, the wine is spilled, and the wineskins are ruined. But they put new wine into new wineskins, and both are preserved." -Matthew 9:17 (also Mark 2:22; Luke 5:37-38) New wine that begins fermenting would burst new or old wineskins. The point here is that the new wine be kept from fermenting; that new wine be preserved. Old wineskins would be impossible to completely clean and the new wine would be infected with the old ferment or yeast. Jesus emphasized new wine and compared it to His ministry and the New Covenant. He contrasted it with the old wine of the law and Old Testament (further discussion in chapter on *Controversial Passages*). Finally, Jesus here calls "wine" (oinos) that which is unfermented. This is another biblical example of nonalcoholic wine being called wine.

Sour Wine or Vinegar is Wine

"Then someone ran and filled a sponge full of sour wine, put it on a reed, and offered it to Him to drink." -Mark 15:36 The word used here is oinos, the general Greek word for wine. So even the last stage of wine, vinegar or ascetic acid is called wine, sour wine. This again shows the broad use of the word wine. And it shows a clear use of the word wine, for wine that is not intoxicating. Dictionaries say that the word vinegar comes from the words meaning sour or sharp wine.

Fruit of the Vine

The cup of the Lord's Supper is never called wine, even though that word is used of both unfermented and fermented wine. Rather, it is called the "cup" or the "fruit of the vine." The most obvious, natural fruit of the vine would be the newly pressed sweet wine or grape juice. It is the direct produce of the vine, rather than the man-made fermented alcoholic wine. (further discussion in chapter on *Controversial Passages*) "But I say to you, I will not drink of this fruit of the vine from now on until that day when I drink it new with you in My Father's kingdom. -Matthew 26:29 (also Mark 14:25; Luke 22:18). The most obvious, natural fruit of the vine is grapes and the fresh pressed juice of the grapes. One day Jesus and His disciples will again drink the fruit of the vine together. It will be new, drug-free wine. It is also interesting that according to this passage, Jesus is today a total abstainer.

Wine of God's Wrath

"He himself shall also drink of the wine of the wrath of God, which is poured out full strength into the cup of His indignation." -Revelation 14:10 Some say while the Greeks and Romans may have mixed their wine with water, they did not do so in Israel. The burden of proof, however, should be on them to explain why everyone else did so except the Jews. This Scripture, nevertheless, speaks of full strength, unmixed, or undiluted wine. Why would it point out full strength wine, if it were already their common practice to drink it undiluted? No one today speaks of drinking undiluted wine, since that is what is always used; to say it would be unnecessary. This verse indicates that, like the surrounding cultures, Israel commonly drank wine mixed with water. Furthermore, this context neither yields to alcoholic or non-alcoholic wine. Again the burden of proof is on them.

Alcoholic Wine is Distinguished from Ordinary Wine

Their wine is the poison of serpents, and the cruel venom of cobras. -Deuteronomy 32:33

The wine of confusion. -Psalm 60:3

Drink the wine of violence. -Proverbs 4:17
Wine is a mocker. -Proverbs 20:1

Bites like a serpent, stings like a viper. -Proverbs 23:32

It is not for kings to drink wine. -Proverbs 31:4

Do not be drunk with wine. -Ephesians 5:18

In contrast, the Bible speaks of new, sweet, non-alcoholic wine:

The wine that makes glad the heart of man. -Psalm 104:15

The vine said *my* wine cheers both God and men. -Judges 9:13

Drink deeply. -Song of Solomon 5:1

New wine in the cluster of grapes contains a blessing. -Isaiah 65:8

Drink the sweet. -Nehemiah 8:10

Makes young women thrive. -Zechariah 9:17

Jesus will drink new wine with His disciples in the Father's kingdom. -Matthew 26:29

Biblical Allusions to Wine and Vineyards

Scripture often refers to wine and vineyards. Obviously they could produce alcoholic wine, that which God condemned. But wine, the nonalcoholic kind, was a staple food and industry. Major products of vineyards included fresh grapes to eat and drink and conserve, dried grapes or raisins, raisin cakes, grape preserves, sweetener and flavoring for cooking, boiled down thick unfermented wine or grape juice, fresh expressed wine, other forms of unfermented wine. It was sometimes even boiled down and dried into a "fruit leather." All of these products had many variations. These products were healthy, nutritious, and drug free. Had grapes and vineyards only been used, or primarily used, for producing a hard drug, would Jesus have compared Himself to a vine, and His followers as branches bringing forth fruit (John 15)? Is our goal as a believer in Jesus to bring forth large amounts of drugs that destroy good judgment? Certainly not.

****

In view of the many passages in the Bible and in ancient literature that speak of nonalcoholic drinks as wine, perhaps we should not be so quick to assume each time we see "wine" in the Bible that it refers to a strong alcoholic drink. People in Bible times were not drenched in drugs.

Sometimes we read passages that refer to wine without clearly explaining whether they were toxic or not. Passages like: "Then Melchizedek king of Salem brought out bread and wine; he was the priest of God Most High." -Genesis 14:18 Or this one: "Ho! Everyone who thirsts, come to the waters; and you who have no money, come, buy and eat. Yes, come, buy wine and milk without money and without price." -Isaiah 55:1 When we read such passages, maybe, just maybe, we should not immediately presuppose drugs.

Based on biblical and ancient evidence, the words for wine were clearly used for un-intoxicating wine and for intoxicating wine. Even for sour wine or vinegar. Wine can include the vine, the grape, the fresh pressed juice, the preserved nonalcoholic juice, the preserved alcoholic juice, and last, sour wine, ascetic acid or vinegar. All are at times referred to as wine. Based on this evidence, perhaps when we see the word "wine" in the Bible, we should assume it is of the un-intoxicating kind, unless the context proves otherwise.

Statements on Un-Intoxicating Wine in Bible Times

Again, **Dr. Moses Stuart's** comment is revealing, "I regard it as established beyond fair contradiction, that it was a very common thing to preserve wine in an unfermented state, and that when thus preserved it was regarded as of a higher and better quality than any other." [255]

Albert Barnes (AD 1798-1870) was a theologian and Presbyterian minister. A graduate of Princeton Theological Seminary and pastor in Morristown, New Jersey and First Presbyterian Church, Philadelphia, Pennsylvania. More than a million volumes of Barnes' Notes had been printed by 1870. "The common wine of Judea was the pure juice of the grape, without any mixture of alcohol, and was harmless. It was the common drink of the people, and did not tend to produce intoxication." [256]

Ferrar Fenton (1832–1920) produced one of the first modern translations of the Bible in 1903 and had a working knowledge of 25 classical, Oriental, and modern languages. He concludes that wine in the Bible, (Hebrew yayin; Greek oinos), "was not confined to an intoxicating liquor made from fruits by alcoholic fermentation of their expressed juices, but more frequently referred to a thick, non-intoxicating Syrup, Conserve, or Jam, produced by boiling, to make them storable as articles of food, exactly as we do at the present day. The only difference being that we store them in jars, bottles, or metal cans, whilst the Ancients laid them up in skin bottles, as Aristotle and Pliny, and other classic writers upon agricultural and household affairs describe." [257]

Fenton adds, "It should never be forgotten that when reading the Bible and the classic pagan writers of 'Wine,' we are seldom dealing with the strongly intoxicating and loaded liquids to which that name is alone attached in the English language, but usually with beverages such as above described. They were as harmless and sober as our own Teas, Coffees, and Cocoas. Had they not been so, the ancient populations would have been perpetually in a more or less state of drunkenness... These facts should never be forgotten when we read of 'Wine' there, - for it was simple fruit syrup, except where especially stated to be of the intoxicating kinds."[258]

Simple Wines of Antiquity

Smith's Dictionary quotes, "The simple wines of antiquity were incomparably less deadly than the stupefying and ardent beverages of our western nations. The wines of antiquity were more like sirups; many of them were not intoxicant; many more intoxicant in a small degree; and all of them, as a rule, taken only when largely diluted with water. They contained, even undiluted, but 4 or 5 percent of alcohol." [259]

Wycliffe Bible Encyclopedia states, "Heb. Yayin 'seems to be used to describe 'all sorts of wine' (Neh. 5:18), from the simple grape juice, or a thickened syrup, to the strongest liquors with which the Israelites were acquainted, the use of which often led to deplorable scenes of drunkenness' (*Fairbairn Imperial Standard Bible Encyclopedia*)" [260]

Adrian Rogers affirms, "Every time the Bible uses the word wine it does not necessarily mean that which is intoxicating. In the Bible there is wine that is intoxicating. In the Bible there is wine that is not intoxicating. And if you don't understand that you are going to be hopelessly confused. You're going to think Jesus perhaps became the original distiller when He turned water into wine. No, not at all." [261]

Jack Graham states, "The Bible talks of two different kinds of wines - one that is permissible and another that is not; one that is similar to grape juice before fermentation and another that is intoxicating. When intoxicating wine is presented in the Bible, it is strongly condemned." [262]

Just as the Bible contains more than one definition and use of words such as God/god, Spirit/spirit, angel, the same is also true of the word wine.

Use of the Word Wine in the English Language

Today the primary use of the English word "wine" refers to fermented wine. But that has not always been the case. If you think a word never changes meaning, just research the word "gay." In earlier days wine was often used in a generic way. While less common, wine is still occasionally used today of unfermented wine or grape juice. This is not just true of wine; many words in today's English language have more than one acceptable meaning.

Early English Dictionaries

And early English Dictionary defines, "Wine - a Liquor made of the Juice of Grapes, or other fruit. Liquor - anything that is liquid; Drink, Juice, Water, &c." [263] An 1800s Encyclopedia, "The juice of the grape, when newly expressed, and before it has begun to ferment, is called must, and, in common language, sweet wine." [264] "Webster defines must (Latin, mustum), 'wine pressed from the grape, but not fermented.'" [265] In a sort of backdoor acknowledgment of unfermented wine, some dictionaries define "must" as "new wine." A 1980 Dictionary defines must as coming from the Latin word "mustum, new wine." [266]

Legal Definitions of Wine

One of the reasons the word wine is not used as much today for non-alcoholic wine is because of governmental and legal regulations. For example, in 2009 a low alcohol winemaker was having trouble with United Kingdom and European Union authorities. It seems they have concluded that if a wine is below about 9% alcohol, it cannot legally be sold as wine. Most wines today have about 12% to 14% alcohol. [267] Dr. Mian N. Riaz, Ph.D, explains, "By law all wine must contain not less than 7% and not more than 24% of alcohol by volume." [268] This is one of the reasons the word "wine" is less frequently used to refer to unfermented or nonalcoholic wine.

Difference in Taste

Another problem for new wine, sweet wine, or what we usually call grape juice is that if it is labeled "wine," drinkers of alcoholic wine get upset because they expect it to taste like alcoholic wine. Unfermented wine tastes very different from fermented wine.

English Hymns

English hymns about the Lord's Supper often refer to wine, even though the churches that sing those hymns use grape juice, or unfermented, nonalcoholic wine for the Lord's Supper. They regularly refer to the cup or fruit of the vine as wine. For example, the *Baptist Hymnal*, LifeWay, 2008 includes the hymns, "Come to the Table," "In Remembrance," and "Jesus, at Your Holy Table." Each of these use the word wine. While it can be argued it is more biblically accurate to use the terms "cup," and "fruit of the vine," this is further evidence that in a denomination that opposes alcohol and uses grape juice for Communion, this Communion cup is still at times called wine.

Non-Alcoholic Wine Aisle

Large wine stores often have an aisle labeled "Non-Alcoholic Wine." This is another example of wine, even today, being used in a nonalcoholic context.

Modern English Translations of the Bible

As discussed earlier in this chapter, modern, as well as older, English translations of the Bible translate what is clearly nonalcoholic wine as, in fact, "wine." See Proverbs 3:10; Isaiah 16:10; 65:8; Joel 2:24 in various English translations. For example, Joel 2:24 unmistakably speaks of new unfermented wine or grape juice (tirosh). Only unfermented wine is pressed out of the grapes in the vat. Yet the KJV, NKJV, ASV, NASB, NLT, ESV, YLT, NIV, HCSB English translations all translate this verse as "wine" or "new wine." This is perhaps the clearest example of the generic use of wine in the today's English language.

CHAPTER SIX:
The Bible Speaks Directly Against Alcohol

Today many are saying that while the Bible condemns drunkenness, it never condemns the moderate use of alcohol. They say that good Christians can be biblically free to enjoy a glass of wine or a beer on the weekends. A close look at the Bible, however, makes it plain that beverage alcohol is condemned, both directly and indirectly.

The Hebrew Bible, or the Old Testament, Directly Condemns Alcohol

Proverbs 23:29-35 Who has woe? Who has sorrow? Who has contentions? Who has complaints? Who has wounds without cause? Who has redness of eyes? 30 Those who linger long at the wine, those who go in search of mixed wine. 31 Do not look on the wine when it is red, when it sparkles in the cup, when it swirls around smoothly; 32 At the last it bites like a serpent, and stings like a viper. 33 Your eyes will see strange things, and your heart will utter perverse things. 34 Yes, you will be like one who lies down in the midst of the sea, or like one who lies at the top of the mast, saying: "They have struck me, but I was not hurt; They have beaten me, but I did not feel it. When shall I awake, that I may seek another drink?

In Bible times they had no word for alcohol. Therefore to describe an alcoholic drink they had to define what it did. The effects of alcoholic wine are given in great detail in the passage by King Solomon.

And Solomon, the smartest man in the world, said not to even look on this kind of wine (v. 31). Unfermented wine does not fit the Proverbs 23 description. You are free to look at it and drink it. That is why God commends some wine (the unfermented kind) and condemns other wine (the fermented, poisonous kind).

Some deflect the force of this commandment by saying, "'Do not look on the wine,' is referring to the drunkard." The context of this passage, however, would disagree. Throughout this chapter, the writer is referring to "my son" and giving him fatherly advice. This command against even looking at alcoholic wine is given to us all, not just the alcoholic. It is also significant that Solomon does not give a detailed description of alcoholic wine, and then say drink this wine moderately. Rather, he says to stay away from it; don't even look at it.

Scripture here describes alcoholic wine and says not to even look at it. In contrast, Scripture never describes alcoholic wine and says drink it.

Proverbs 20:1 *Wine is a mocker, strong drink is a brawler, and whoever is led astray by it is not wise.*

Notice it says "Wine IS a mocker." This verse is obviously speaking of the alcoholic kind of wine. This wine in and of itself is condemned by God. It is not the using of wine in excess that is condemned. The wine itself is a mocker. This alcoholic wine (and similar drink, strong drink, or alcoholic shekar) itself is denounced. Non-alcoholic wine or grape juice is not a mocker. Intoxicating wine, however, is a strong, hard drug. It is a mocker and brawler and those deceived into thinking otherwise are not wise. Those who play with the fire of hard drugs are foolish.

In these two passages God unambiguously conveys that beverage alcohol is wrong; it is not wise.

Sampling from several commenters on these passages we will see the agreement.

Jim Richards affirms, "Proverbs 23:31 makes perhaps the strongest case for a command to not use alcoholic wine. While some Hebrew scholars point to the difficulty of the translation to English, there is little dispute that the verse gives a prohibition to even look on fermented wine in this context. Proverbs 20:1 is another verse condemning the use of wine and strong drink." [269]

Barry Creamer colorfully explains, "But it is easy to see passages like Proverbs 23 making the obvious point that 'abuse' is the ultimate problem and that therefore total abstinence is the solution. Like many illicit drugs, the essential attributes of alcohol are its intrinsically (not simply subjectively psychologically) addictive, directly mind-altering characteristics. Those characteristics are what lead to the Proverb's point not simply to avoid drunkenness, but to avoid any drink at all, and even to avoid those who drink. To illustrate, you will not manage to make any argument for the moderate consumption of alcohol which does not also apply to the moderate consumption of marijuana–if it were legal. Do you seriously think Jesus would have sat down for a joint with his homies?" [270]

Charles Wesley Ewing says, "'Wine is a mocker, strong drink is raging: and whosoever is deceived thereby is not wise' (Proverbs 20:1). Sweet unfermented wine is no mocker. It is not raging. It does not deceive the user. Fermented wine is the mocker, rager, deceiver." [271]

Adrian Rogers sums up, "These Scriptures (Proverbs 20:1; 23:29-32) tell us, I believe plainly and clearly, that the Christian position so far as beverage alcohol is concerned is total abstinence. The Bible says we are not to look upon it, we are not to desire it when it is fermented. Beverage alcohol is America's most dangerous drug." [272] A Study Bible concludes, "The warning is to avoid even looking at the alluring wine (v. 31)." [273]

Proverbs 23:29-35 and Proverbs 20:1 are two Old Testament passages that clearly condemn drinking. They plainly teach abstinence from intoxicating wine.

The New Testament Portion of the Bible Directly Condemns Alcohol

The New Testament also directly condemns the use of alcohol.

1 Thessalonians 5:6-8 Therefore let us not sleep, as others do, but let us watch and be sober.

For those who sleep, sleep at night, and those who get drunk are drunk at night. But let us who are of the day be sober.

1 Thessalonians 5:6-8 tells us to be sober. It even contrasts drunkenness with sobriety. Any amount of alcohol makes us less than sober. Moderately drink, and you will be moderately sober. Ask Alcoholics Anonymous; their definition of "sober" allows absolutely no alcohol.

Other verses also emphasize the idea of sobriety.

1 Peter 1:13 Therefore gird up the loins of your mind, be sober, and rest your hope fully upon the grace that is to be brought to you at the revelation of Jesus Christ.

1 Peter 5:8 Be sober, be vigilant; because your adversary the devil walks about like a roaring lion, seeking whom he may devour.

If God commands you to be sober, how can you drink an intoxicating drink without violating that command? How can a believer take a mind-altering drug for his amusement and recreation, and claim to be sober?

Some have claimed this word for sober has nothing to do with drinking. While the word sober would speak to more than just drinking, how can a command to be sober not include drinking an intoxicating beverage? Whatever else sober implies, it most certainly would also cover the recreational use of a hard drug, whether legal or not.

2 Timothy 4:5 But you be watchful in all things, endure afflictions, do the work of an evangelist, fulfill your ministry. The word "watchful" is the same Greek word (nepho) translated elsewhere as "sober." The Apostle Paul was directly commanding the young preacher Timothy to be sober. Timothy took that command so seriously he drank only water.

He did not even drink unfermented wine or grape juice until advised to do so by Paul for medicinal purposes.

Variations of words for sober (Greek - nepho, sophron) are also used in the following verses: Romans 12:3; 2 Corinthians 5:13; 1 Timothy 3:2; Titus 1:8; 2:2; 2:4; 2:5; 2:6; 1 Timothy 3:11; 1 Peter 4:7. While other things are also implied, you cannot at the same time "think soberly" (Romans 12:3) and indulge in intoxicating drink.

W.E. Vine, a Greek scholar says that "Nepho (sober) signifies to be free from the influence of intoxicants; in the NT, metaphorically, it does not in itself imply watchfulness, but is used in association with it." [274] The Greek word nepho (or nephalios) is the word used for sober in 1 Thessalonians 5:6,8; 2 Timothy 4:5; 1 Peter 1:13; 4:7; 1 Timothy 3:11; Titus 2:2. Vine also defines a similar word, "Sophron - denotes of sound mind; hence, self-controlled, soberminded." [275] Once again, you cannot drink and be self-controlled or sober minded. The first thing alcohol does is affect your judgment.

John MacArthur explains, "Nephalios (temperate) literally means 'wineless,' or 'unmixed with wine.'" [276]

Kittle's Theological Dictionary states: "It is compelling that the Greek word for sober (self-control) in numerous passages (1 Thess 5:6; 1 Thess 5:8; 1 Tim. 3:2; 1 Tim. 3:11; 1 Peter 1:13; 1 Peter 4:7; 1 Peter 5:8) is a word that also means, "holding no wine" [277]

The Bible clearly, directly speaks against using the drug of beverage alcohol.

CHAPTER SEVEN:
Biblical Teaching Condemns Alcohol

Not only does the Bible *directly* teach abstinence from beverage alcohol, but biblical *principles* also condemn the use of alcohol. Remember that beverage alcohol is the recreational use of a harmful, mind-altering drug. God's Word in no way condones such a practice. The same could be said for any unnecessary use of any drug. Those who favor the moderate use of alcohol are really advocating the use of a hard drug simply for recreational or pleasurable purposes. There are a host of biblical principles against such a practice. When it comes to alcohol and other drugs: Don't start. If you have started, stop. Stay away from them. Avoid them as you would a venomous snake.

Biblical Principles that Condemn Alcohol

1. The Bible says those deceived by wine are not wise (Proverbs 20:1).

We are commanded to be wise (Ephesians 5:15; etc.).

2. The Bible teaches us to guard our influence and not to lead others astray.

You may be able to hold your liquor. There are still at least two problems with that. First, you are supporting an evil industry that has brought untold heartache to the world. Second, someone else will look at you and say, "That is the best man I know. If he can drink then so can I." And that may be someone whose life will be ruined by drink.

"Having lived and served as a missionary in several countries where the lives of wives and children have been devastated by an alcoholic husband and father have taught me that the farther one stays from alcoholic drink the better one will be in providing a good example and in helping those who are enslaved by alcoholism." **Dr. Daniel R. Sanchez** (DMin, PhD), Professor of Missions, Southwestern Baptist Theological Seminary, Fort Worth, Texas.

3. Your body is the temple of the Holy Spirit (1 Corinthians 3:16; 6:19-20).

Alcohol weakens and destroys that body.

4. The Bible says you are to love God with all your mind (Matthew 22:37).

Our minds are altered and damaged by alcohol. Every drink kills brain cells and dulls your judgment. When it comes to beverage alcohol, its very use is abuse. We should love and serve God with clear minds.

5. The law of love, as A. T. Robertson called it, teaches us not to drink (Romans 14:19, 21; 1 Corinthians 8:9).

We are to be willing to deny ourselves, not abuse our liberties, out of love for our brothers in Christ; and out of love for those who are non-believers. Don't be a stumbling block to others.

6. Scripture proclaims us kings and priests (1 Peter 2:5-9; Revelation 1:6; 5:10).

Kings are not to drink lest they pervert justice (Proverbs 31:4-5). Notice there is no prescription for kings to drink moderately. Rather, they are not to drink wine, period. Kings are admonished not to drink wine in any amount.

7. Priests were commanded not to drink during their duties so that they could distinguish between what is holy and unholy (Leviticus 10:8-10).

A Christian today is a priest (1 Peter 2:5-9), and he needs to be able to distinguish between what is holy and unholy. Alcohol removes your ability to do so. By the way, commanding a priest not to drink during worship does not equal a command that he drink outside of worship. Nor does it equal permission to drink at other times. Is there ever a time when a believer does not need to be able to distinguish between what is holy and unholy?

Ernest Gordon states, "One should at this point recall Ezekiel's words (44:21), 'Neither shall any priest drink wine, when they enter into the inner court,' and ponder Christ's words (Matt. 6:6), 'But thou, when thou prayest, enter into thy closet, and when thou hast shut thy door, pray to thy Father which is in secret.' Here is the priestly ministry of the believer of the new dispensation...Wine and prayer are incompatible things." [278]

8. God commended the Rechabites for not drinking wine (Jeremiah 35).

God highly praised them for keeping their commitment through the years to stay away from wine.

9. Don't abuse your Christian liberty (1 Corinthians 8:9; 10:23).

Judge Paul Pressler puts it, "The upcoming generations need to know the havoc brought on our society and upon individuals by the use of alcohol. If we use it ourselves, we recommend its use to others. A Christian should not exercise his freedom to put himself and others at such a risk." [279]

10. The Bible often gives the appalling results of alcohol.

Scripture clearly relates the terrible consequences of Noah, Lot, and others getting drunk. It never commands anyone to drink of this alcoholic wine.

11. Is it biblical for a believer to support the alcohol industry that has wrecked so many homes and lives?

Is it biblically permissible to support them with your God given resources?

12. Drinking is expensive.

God gave you the ability to get wealth. By not drinking you can save a lot of money that you can use for more noble purposes.

During a news segment, a woman asked of her response to the recession of 2009 replied, "I'm no longer buying wine." Others have said, "I wish I had the money I spent in the past on drinking." Alcohol is expensive. You can save yourself a world of problems, and money, by leaving it alone.

13. Biblical wisdom and truth would compel us to recognize the incredible damage alcohol does to society.

Alcohol contributes greatly to traffic accidents and deaths, unwanted pregnancies, fetal alcohol syndrome, the spread of sexually transmitted disease, all types of criminal behavior, cirrhosis of the liver, destruction of brain cells, addiction, breakup of homes. A Christian should recognize this and stay away.

14. About one out of nine drinkers becomes a problem drinker.

Never take that first drink and I guarantee you will never become an alcoholic. With those odds in mind, would biblical principles allow you to play "Russian Roulette" with your life and the lives of those you love?

15. From the overall teaching of the Bible, do you really believe God condones the recreational use of a mind altering, dangerous drug?

Whether it is alcohol, marijuana, cocaine, etc., the answer is obviously, no.

16. Countless lives have been saved from ruin by teaching abstinence from alcohol.

What is the worst that can happen to you by not drinking? Abstinence works every time it is used. Not drinking is safe, and it is wise. You don't have to drink; you can graciously say no. You can help save the lives of others by teaching abstinence.

17. Addiction to alcohol and other drugs is a serious problem.

The first drink of alcohol is often a gateway to addition and to other dangerous drugs. Jesus said, "Whoever commits sin is a slave to sin" (John 8:34). With beverage alcohol, it is easier to not start, than to start and then struggle to stop.

God's plan A is for us not to sin in the first place (1 John 2:1). As it is, you will have plenty of problems in life. Abstinence can save you from many of the self-induced ones. You can save yourself and others a world of heartache by just staying away from alcohol.

18. The Bible teaches self-denial, not selfish gratification.

Jesus Christ said, "If anyone desires to come after Me, let him deny himself, and take up his cross, and follow Me" (Matthew 16:24).

When it comes to biblical principles, alcohol is kind of like the issue of slavery. The Bible does not directly say, "Thou shalt not own a slave." But clear biblical principles certainly teach against slavery. Whether or not you accept that the Bible directly speaks against alcohol, the Bible clearly teaches against it.

Bible Verses Directly or Indirectly Teaching Against Alcohol and Drugs.

Listed here are a few Bible verses that directly and indirectly present the case for abstinence from beverage alcohol and other destructive drugs.

Leviticus 10:9; Proverbs 20:1; 23:29-35; Isaiah 28:7; Jeremiah 35; Daniel 1:8; Habakkuk. 2:15; Romans 14:19, 21; 1 Corinthians 3:16; 6:19-20; 8:9; Ephesians 5:15; 1 Timothy 3:3; Titus 1:7; 1 Peter 1:13; 5:8; etc.

The Cost of Alcohol

Alcohol causes 79,000 deaths a year in the United States (as of 2010). Even in small amounts it slows reaction time and impairs judgment and coordination leading to accidents of all kinds. Alcohol takes away good judgment and inhibitions causing men to do things they would never do in their right minds. It breeds aggression, crime, suicide, immorality, unwanted pregnancy and sexually transmitted disease. It increases the risk of cancer and causes mental and birth defects in unborn children. Women who drink during pregnancy are more likely to have the child die from Sudden Infant Death Syndrome. Drinking increases a woman's risk of miscarriage. Alcohol causes cirrhosis of the liver and other liver diseases. Alcohol leads to memory loss and shrinkage of the brain. Alcohol damages heart muscle and causes other heart disease. It makes men more likely to commit sexual assault and women more likely to be sexually assaulted. It causes dangerous reactions with legitimate over the counter and prescription drugs. Alcohol worsens other medical conditions and damages skill, coordination, alertness, and judgment. In turn it leads to unintentional injuries, including traffic injuries, falls, drownings, burns, and unintentional firearm injuries. Brings about child abuse and neglect.Psychiatric and social problems. Harms you and those you love. Causes stomach problems. Alcohol is addictive. It destroys homes. Produces spiritual and moral problems. Alcohol takes the problems you already have, and makes them worse. It harms our influence and leads others down the wrong road. Beverage alcohol is poison. Not drinking causes none of these risks. With this in mind, is it ever wise to drink?

Chapter Eight:
Controversial Biblical Passages Dealing with Wine

Jesus Turned Water into Wine

John 2:1-11 On the third day there was a wedding in Cana of Galilee, and the mother of Jesus was there. 2 Now both Jesus and His disciples were invited to the wedding. 3 And when they ran out of wine, the mother of Jesus said to Him, "They have no wine."4 Jesus said to her, "Woman, what does your concern have to do with Me? My hour has not yet come." 5 His mother said to the servants, "Whatever He says to you, do *it.*" 6 Now there were set there six waterpots of stone, according to the manner of purification of the Jews, containing twenty or thirty gallons apiece.

7 Jesus said to them, "Fill the waterpots with water." And they filled them up to the brim. 8 And He said to them, "Draw *some* out now, and take *it* to the master of the feast." And they took *it.*9 When the master of the feast had tasted the water that was made wine, and did not know where it came from (but the servants who had drawn the water knew), the master of the feast called the bridegroom. 10 And he said to him, "Every man at the beginning sets out the good wine, and when the *guests* have well drunk, then the inferior. You have kept the good wine until now!" 11 This beginning of signs Jesus did in Cana of Galilee, and manifested His glory; and His disciples believed in Him.

Scholarly professors have made fun of the faithful Sunday School teacher who taught that Jesus did not turn water into intoxicating wine.

Based on the evidence, however, it is likely that the esteemed professor is the one who is wrong. But by no means do all learned professors believe Jesus turned water into a hard drug; many know better.

Some actually believe this was a drunken wedding feast. After all they had drunk all the wine (v. 3). The master of the feast said the guests had well drunk (v. 10). The NIV even says, "too much to drink." They believe that Jesus' contribution to this drunken feast was to create over 120 gallons of more intoxicating wine. Though that is a bartender's dream, it is in no way characteristic of the sinless, holy Lord Jesus Christ of Scripture.

Remember that wine in the Bible and in ancient times referred to both fermented and unfermented wine. The Greek word used in John 2 is oinos. Clearly un-intoxicating wine in Proverbs 3:10 and Isaiah 16:10 was translated with oinos by the Jewish scholars of the Septuagint. So whether you believe the wine (oinos) Jesus made was alcoholic wine or nonalcoholic wine is your interpretation. It is not that you are simply taking the Bible for what it says. The Bible never says Jesus turned water into alcoholic wine. The Bible never says Jesus or His disciples drank fermented wine.

Reasons to Believe Jesus Did Not Create Intoxicating Wine in John 2

1. The Greek word oinos, used here for wine, is a generic word used of both intoxicating and un-intoxicating wine.

2. Preventing fermentation would not have been a problem here; but the ancients knew well multiple ways of keeping and preserving nonalcoholic wine. Nonalcoholic wine was a common drink.

3. Fermented wine involves a process of rotting, decomposing the wine and allowing it to become intoxicating. Why would Jesus make something clean, fresh, and pure, and then put it through such a decomposing, fermenting process? Why

would He have taken the sweetness and nutrition away? Why not let this new wine reach and retain its peak of perfection? Jesus did not create rotten bread and decomposed fish at the feeding of the 5,000.

4. Through an amazing year long process rain comes down from above, the roots take up moisture and nutrients, leaves take in the sunshine, blooms form, and slowly grapes form, grow, and ripen. In short, water turns to wine. God turns water into wine every year in every vineyard; in this instance Jesus just sped up the process.

5. God makes the grape and the pure juice; man corrupts it into a drug that will cause a person to do things he would never do in his right mind.

6. This miracle revealed Jesus' glory (John 2:11). Making copious amounts of a dangerous drug would not manifest His glory.

7. Mary, the mother of Jesus was present at this wedding feast. Mary was not sinless (Luke 1:47; Romans 3:23), but she was a godly woman and great example. Would she have been at a drunken party? Would she have encouraged others to assist her Son in making huge quantities of alcohol?

8. Children would have been present at this wedding; another good reason to exclude alcoholic wine.

9. No hint of drunkenness is mentioned in this story. The only thing that got out of hand was the fact that they ran out of wine. Apparently the guests were well-behaved, just thirsty.

10. Jesus and His disciples were invited to this party (John 2:2). That alone indicates this was a joyous but drug free, sober celebration. Christians often have such parties today. Fun, joy-filled gatherings that leave no regrets. Non-Christians also often have drug-free parties.

11. Drinking great quantities of wine, with no hint of a problem, strongly indicates that this wedding, like so many today, was free of alcohol. They drank all the wine the wedding party

provided; then Jesus provided much more. Yet with no problems whatsoever. That signifies soft wine, not hard wine.

12. Jesus made a huge amount of wine. He made at least 120 gallons. He gave no restriction about its use. Apparently they were to rejoice and drink till they could drink no more. He could not have righteously done this with alcoholic wine.

13. This was Jesus' first miracle, a sign of the new covenant. The old, inferior covenant was drawing to a close; the New Testament was present. The new covenant (Hebrews 9:15) is much better represented by wine that is new, sweet, morally good, rather than by that which is decomposed and will make a man intoxicated. Jesus even compared His ministry with new wine, and contrasted it with the old wine of the old covenant.

14. Scripture never says the guests were intoxicated; to infer that is an interpretation.

15. Scripture never says Jesus made intoxicating wine. Scripture never says Jesus drank intoxicating wine.

16. Morally, the "good wine" is that which is drug-free, safe, and healthy.

17. God often saves the best for last. You can be assured that as the guests returned home they were not guilty of drunk walking, drunk donkey riding, or drunk chariot racing. No drunk men went home that day and beat their wives and children.

18. Some point out this wine did not have time to ferment. It was probably used immediately. Wine has to have time and the right temperature range to ferment. Yes, Jesus could have instantly fermented the new wine, but why?

19. This was obviously "new wine." New wine was commonly understood as being unfermented.

20. The burden of proof should lie on those who believe Jesus created great amounts of intoxicating wine. Since wine referred to both un-intoxicating and intoxicating wine, why do you choose to interpret this miracle as Jesus creating a dangerous drug?

Reasons to Believe Jesus Did Not Create Intoxicating Wine in the Entire Context of Scripture

1. If Jesus did create intoxicating wine at Cana, it obviously endorses the use of alcoholic wine and other hard drugs for recreational use. This wedding feast was in no way a medicinal use of wine. Scripture does not endorse such drug use.

2. Jesus would have violated the biblical mandates against intoxicating wine in Proverbs 20:1 and 23:29-35. He would have violated the later New Testament commands to be sober. Peter, present at this miracle at Cana, would later warn us to be sober (1 Peter 5:8). The first drink of alcohol ends your sobriety.

3. Assuming Jesus drank that wine, as our Great High Priest and King of Kings, He would have violated Proverbs 31:4-5.

4. If Jesus had created intoxicating wine, He would have violated the many biblical principles against the recreational use of a mind-altering drug.

5. Jesus would have violated the spirit, if not the letter of the law, of Habakkuk 2:15 against giving drink to your neighbor and making him drunk.

6. That Jesus made great quantities of intoxicating wine, a hard drug, is incompatible with all we know of His character as the holy, loving, sinless, all-wise God (Hebrews 4:15; 9:28; Acts 4:30; John 1:1; Titus 2:13; Revelation 1:8).

7. Saying "most serve the best first, but you have saved the best for last," in no way implies this wine had to be alcoholic. At gatherings today good Christian people often serve the best first. That does not mean what they are serving has to be alcoholic.

8. Jesus first miracle was creating new un-intoxicating wine. His ministry was compared to new wine. And His last mention of wine is when He will drink it new in the kingdom (Matthew 26:29; Mark 14:25; Luke 22:18). All these references point to new, drug-free wine.

143

Objections

1. If it Was Not Fermented, Why Did the Master of the Feast Say Jesus' Wine Was the Best?

When you think of it, isn't that a foolish question? If Jesus directly and miraculously made anything, wouldn't it be the best? It is also going to be morally the best. Second, references have been given previously about the common wine in Jesus' day. The common wine in Israel was either un-intoxicating or very lightly alcoholic. This was not necessarily a comparison with alcoholic wine, for most likely the wedding feast celebrated with unfermented wine. But whatever the first wine, the un-intoxicating pure, new wine made by Jesus would have been superior. His wine was new, sweet, and perfect. Third, most everyone, upon their first taste of wine, would prefer sweet nonalcoholic wine or grape juice over the bitter taste of alcoholic wine - unless they are looking for the drug effect of alcoholic wine. Fourth, that this wine was good wine, speaks not only of its quality of taste, but also a morally good wine.

2. But Doesn't the Bible Say the Guests at the Wedding were Drunk?

No, it does not. The guests had drunk the wine available at the wedding feast. That wine was very likely good, fresh, sweet un-intoxicating wine. Second, the master of the feast did not say the guests at that feast had drunk freely; his was only a general statement. Third, the word used for "well drunk" could mean intoxicated, but did not always mean such. It is a word that simply meant filled. You can be filled with alcohol and therefore be drunk, or you can be filled with grape juice or a soft drink. Fourth, this was not a drunken party, but if it were, that would be more reason not to believe Jesus would make something that would make them even more intoxicated.

The Nature of Oinos

Dr. Paige Patterson writes, "In Jesus' miracle at Cana of Galilee (John 2:1-11), one can neither affirm with certainty that Jesus turned the water into a non-intoxicating wine nor that He drank no wine Himself." [280] Similarly, Dr. Richards attests that, "We do not know if the "oinos" was fermented. The pure blood of the grape would have had no alcoholic content. One interpretation is just as valid as the other." [281] Finally we will note what author Orin Whitmore writes, "It is true that the Greek word here translated wine is the generic term, and we cannot tell by the term itself just what kind of wine it was - intoxicating or non-intoxicating. So far as the word itself is concerned, there is nothing in the word to indicate that it was intoxicating, any more than there is to show that it was non-intoxicating, hence there is no evidence to be adduced from the word itself to prove that the wine made out of water was intoxicating." [282]

As we can see in contrast to many false views, oinos itself could be interpreted either way - as fermented or unfermented wine. Therefore, either view on the nature of the wine Jesus made, is an interpretation, not just taking the Bible for what it says. Thus, the charge of reading into the text can be equally applied to both sides, at least at this juncture. The arguments demand more careful study, as the following quotes reveal.

Contrary to some, not all wine of Bible times was fermented; much of the wine is safe, un-intoxicating, non-alcoholic. To drive home this point Orin B. Whitmore said, "Is all wine 'a mocker'? Then it was 'a mocker' that Jesus made for the guests at the wedding feast of Cana, and 'a mocker' which Jesus introduced to His disciples at the Passover table, and bade them to drink. Does all wine 'bite like a serpent' and 'sting like an adder'? Then Jesus made wine for the guests at Cana with the 'bite of a serpent' and the 'sting of an adder' in it. Do you believe it? No, a thousand times no! Did Jesus give to His disciples a cup in which were the 'bite of the serpent and the sting of the adder,' and tell them that cup contained that which represented His blood, His life-giving blood - shed for the remission on their sins? Do you believe it? No..." [283]

Leon C. Field speaks, "It must also be observed that the adjective used to describe the wine made by Christ is not agathos, good, simply, but kalos, that which is morally excellent or befitting. The term is suggestive of Theophrastus' characterization of unintoxicating wine as moral (nthikos) wine." [284] Paige Patterson explains, "From a standpoint of logic, the "oinos" that Jesus produced was more likely pure, rather than fermented, grape juice, since that which comes from the Creator's hand is inevitably pure. Also, there was no time for fermentation to take place subsequent to the miracle. Furthermore, the ancients always acknowledged that the best "oinos" was the unfermented "oinos," i.e., that which came from the initial mixing of the grapes." [285]

Patterson continues, "The governor of the feast obviously was able to identify "good wine" by tasting it, indicating that there was no intoxication on his part. On the other hand, by the governor's own testimony, by the last stages of such a feast participants generally had their senses sufficiently dulled so that they could not differentiate between good and bad wine. Was this feast different? Is this why Jesus agreed to attend?" [286]

Theologians on Turning Water to Wine

Chrysostom, born in Antioch in AD 347, was a Church Father, a theologian, and known as an eloquent preacher. Speaking of Jesus turning water into wine, "Now indeed making plain that it is He who changes into wine the water in the vines and the rain drawn up by the roots, He produced instantly at the wedding feast that which is formed in the plant during a long course of time." [287]

The church Father, Augustine, says "He that had made wine that day in those six water-pots does the same every year in the vines. For as what the servants put in the water-pots was changed into wine by the operation of the Lord, just so what the clouds pour forth is changed into wine by the operation of the same law." [288]

H. A. Ironside (1876-1951), born in Toronto, Canada, beloved pastor of Moody Memorial Church, Chicago, Illinois and author of numerous expositional Bible commentaries, has this to say, "It was a wonderful miracle, and yet, after all, it was just a duplication of what our Lord Jesus Christ has been doing for millenniums on ten thousand hillsides, changing water into wine." [289]

F. B. Meyer (AD 1840-1925) was born and ministered in Great Britain. Educated at London University and Regent's Park Baptist College, he helped launch the British ministry of a young D. L. Moody. Pastor, popular speaker and author, he states, "The sin of drunkenness was not the sin of Palestine, as it is of London; and therefore did not require the special methods of prevention which the principles of His Gospel now lead us to adopt. And we must remember that the light wines of the Galilean vintage were very different from the brandied intoxicants with which we are too familiar. But this is the interesting point: that we see compressed into a single flash the same power that works throughout the winelands every summer, transforming the dew and rain into the juices that redden the drooping clusters of the vines." [290]

F.F. Bruce says, "Jesus' action was, in C. S. Lewis's terminology, a 'miracle of the old creation': the Creator who, year by year, turns water into wine, so to speak, by a natural process, on this occasion speeds up the process and attains the same end." [291] C. S. Lewis was not an abstainer, but he and Bruce were right on this point and followed in the line of Chrysostom, Augustine, and many others.

George Whitefield Samson was a graduate of Brown University and Newton Theological Seminary. He was a highly educated, respected Baptist pastor and president of George Washington University and later at Rutgers. The author of several books shares "Wine is nothing else than water, having in solution the sugar, spice and gluten which form grape-juice; and the product which, in the natural development, is slowly made, was by Christ's interposition instantaneously formed." [292]

Adam Clarke (AD 1760-1832) was a British Methodist Bible scholar. His commentary was a primary theological resource for many years. Here are his comments on John 2:10. "'The good wine until now.' That which our Lord now made being perfectly pure, and highly nutritive!... 'But did not our Lord by this miracle minister to vice, by producing an excess of inebriating liquor?' No; for the following reasons: 1. The company was a select and holy company, where no excess could be permitted. And, 2. Our Lord does not appear to have furnished any extra quantity, but only what was necessary. 'But it is intimated in the text that the guests were nearly intoxicated before this miraculous addition to their wine took place; for the evangelist says, otan mequsqwsi, when they have become intoxicated.' I answer: 1. It is not intimated, even in the most indirect manner, that these guests were at all intoxicated. 2. The words are not spoken of the persons at that wedding at all: the governor of the feast only states that such was the common custom at feasts of this nature; without intimating that any such custom prevailed there. 3. The original word bears a widely different meaning from that which the objection forces upon it. The verbs mequskw and mequw, from mequ, wine, which, from meta quein, to drink after sacrificing, signify not only to inebriate, but to take wine, to drink wine, to drink enough: and in this sense the verb is evidently used in the Septuagint, Genesis 43:34; So v. 1; 1 Macc. xvi. 16; Haggai i. 6; Ecclus. i. 16. And the Prophet Isaiah, Isa. 58:11, speaking of the abundant blessings of the godly, compares them to a watered garden, which the Septuagint translate, wv khpov mequwn, by which is certainly understood, not a garden drowned with water, but one sufficiently saturated with it, not having one drop too much, nor too little." [293]

Dr. Daniel D. Whedon expressed it, "And no doubt Jesus, like the God of nature, created, not the alcohol, which is the poison produced by the putrefying corpse of the dead grape, but the fresh, living, innocent fluid." [294]

Albert Barnes (AD 1798-1870) was a theologian and Presbyterian minister. He was a graduate of Princeton Theological Seminary and pastor in Morristown, New Jersey and First Presbyterian Church, Philadelphia, Pennsylvania. More than a million volumes of *Barnes' Notes* had been printed by 1870. Here is his note on John 2:10. "The good wine. ..Pliny, Plutarch, and Horace describe wine as good, or mention that as the best wine, which was harmless or innocent - poculo vini innocentis. The most useful wine - utilissimum vinum - was that which had little strength; and the most wholesome wine - saluberrimum vinum - was that which had not been adulterated by 'the addition of anything to the must or juice.' Pliny expressly says that a 'good wine' was one that was destitute of spirit (lib. Iv c. 13). It should not be assumed, therefore, that the good wine was stronger than the other; it is rather to be presumed that it was milder. The wine referred to here was doubtless such as was commonly drunk in Palestine. That was the pure juice of the grape. It was not brandied wine, nor drugged wine, nor wine compounded of various substances, such as we drink in this land. The common wine drunk in Palestine was that which was the simple juice of the grape. We use the word wine now to denote the kind of liquid which passes under that name in this country - always containing a considerable portion of alcohol - not only the alcohol produced by fermentation, but alcohol added to keep it or make it stronger. But we have no right to take that sense of the word, and go with it to the interpretation of the Scriptures. We should endeavor to place ourselves in the exact circumstances of those times, ascertain precisely what idea the word would convey to those who used it then, and apply that sense to the word in the interpretation of the Bible; and there is not the slightest evidence that the word so used would have conveyed any idea but that of the pure juice of the grape, nor the slightest circumstance mentioned in this account that would not be fully met by such a supposition." [295]

Barnes adds, "No man should adduce this instance in favour of drinking wine unless he can prove that the wine made in the 'waterpots' of Cana was just like the wine which he proposes to drink. The Saviour's example may be always pleaded just as it was; but it is a matter of obvious and simple justice that we should find out exactly what the example was before we plead it." [296]

The Word "Methuo," Drunk or Filled?

Jim Anderson defines Methe / Methuo - "In most cases this word would refer to the consumption of intoxicants. But it also may be used to refer to the profuse drinking of a non-intoxicant." [297] Methuo is used in John 2 and 1 Corinthians 11 in a possibly non-intoxicating way. Usually translated "drunk," but sometimes signifies being "filled" or "satiated." Similarly, Barnes acknowledges in his note on John 2:10: "This word does not of necessity mean that they were intoxicated, though it is usually employed in that sense. It may mean when they have drunk sufficient, or to satiety..." [299]

Dr. J. Vernon McGee records, "It is my firm conviction that the Lord Jesus did not make an intoxicating drink at the wedding feast of Cana of Galilee. Anyone who attempts to make of Him a bootlegger is ridiculous and is doing absolutely an injustice. Folk like to present the argument that in the warm climate of Israel all one had to do was to put grape juice in a wine skin and in time it would ferment. Yes, but in the miracle at Cana, the Lord Jesus started out with *water*, and in the matter of a few seconds He had 'wine.' My friend, it didn't have a chance to ferment. And we must remember that the wedding in Cana was a religious service, and everything that had to do with leaven (which is fermentation) was forbidden. This is the reason that at the time of the Passover and the institution of the Lord's Supper the wine could not have been fermented. Fermentation is the working of leaven, and leaven was strictly forbidden in bread and in everything else. The bread and drink could not have been leavened. Intoxicants are condemned in the Word of God [Proverbs 20:1]." [300] McGee was a graduate of Columbia Theological Seminary and Dallas Theological Seminary (Th.D.).

Commenting on the same story, Dr. John R. Rice says, "There is no reason to suppose that the wine which Jesus made at the wedding in John 2 was intoxicating wine." [301]

Turning again to Dr. Teachout we find, "The text implies that the beverage which the Lord made out of water in the wedding emergency was given to guests who had already had much oinos to drink (v. 10). These guests would thus have already become intoxicated if they were drinking [alcoholic] wine rather than grape juice. Then, if Jesus Christ had given them even more wine (instead of simply replenishing their exhausted supply of grape juice) He would have deliberately caused the marriage guests to become even more drunk. Since all Scripture is united in condemning drunkenness as a sin, this kind of harmful miracle is an impossibility for the One 'who knew no sin.' Instead, the text indicates the surprise of the master of the feast at the sparkling freshness of the good (kalon) grape juice (oinos)." [302]

Pastor O. O. Irvin of Kingsville, TX was challenged about Jesus turning water into wine. Irvin replied, "That's good stuff. Anytime you can get wine that has been made out of water, drink all you want." [303]

Was Jesus A Wine Drinker?

The Bible never says Jesus or His disciples drank intoxicating wine. Rather, there is strong reason to believe they did not drink. Jesus was accused by His enemies of being a wine-bibber and a glutton (Matthew 11:18-19). They also accused Him of having a demon (John 10:20-21). If you believe Jesus drank because of this accusation, do you also believe He had a demon? Why would a follower of Jesus believe these false, bizarre accusations? Jesus ate and drank with the common people. Pastors do so today; that does not mean they drink alcoholic beverages. There is strong indication that Jesus and His disciples did not drink because of the biblical admonitions against drinking.

Those mocking the disciples on the day of Pentecost accused them of being drunk on new, sweet wine. This is the wine Aristotle said would not inebriate. This is the wine Plato said he was sending to children. The indication is the enemies of the disciples knew they did not drink strong wine and mocked them as being drunk on a soft drink. This derision was really a back handed compliment. It acknowledged they did not drink intoxicating wine.

Why would the Apostle Paul tell a pastor / bishop not to drink if Jesus and His disciples drank? Is Paul telling preachers to set a better example than Jesus? Rather, Paul is just upholding the standards set by Jesus and His Apostles. Furthermore, Proverbs 31:4 says it is not for kings to drink wine. Jesus is the King of kings and Lord of lords (Revelation 19:16). Why would Jesus, as King and with His sinless nature, go against this command?

Jesus emphasized new wine throughout His ministry. The old inferior covenant was drawing to a close, and the new covenant had appeared. Jesus compared His ministry with new wine (un-intoxicating wine), and contrasted it with the old wine of the old covenant. Jesus reproved those who preferred that old wine. He said "I am the true vine, you are the branches." A vine and branches bring forth fresh, unfermented fruit and juice. Jesus and Scripture refer to the Lord's Supper with the terms "cup" and "fruit of the vine." Fruit of the vine most naturally refers to the new fresh product of the vine. Jesus said He would not drink the fruit of the vine again until he drank it new in the Kingdom. The Bible never says Jesus made or drank alcoholic wine. Jesus lived a sinless, holy life. He was the perfect influence and example. Scripture condemns intoxicating wine and emphasizes sobriety. Sum all this up, and it is clear that Jesus did not partake of intoxicating wine. For Him to have made or drank intoxicating wine would have contradicted His life and ministry.

Theologians Weigh in on the Answer

Paige Patterson convincingly argues, "The accusation that Jesus, in contrast to John, was a socialite, a glutton, and a winebibber is manifestly void of foundation (Matthew 11:19; Luke 7:34). Because Jesus enjoyed social contacts and openly mingled with the people, some assumed that He had a propensity for food and drink. If Jesus had been a winebibber, He must have also been guilty of gluttony, which is clearly identified as a sin. In fact, Jesus was neither, and again there is no evidence that He drank "oinos" or anything other than the fresh, natural fruit of the vine." [304]

Jim Richards asserts, "Jesus was not a winebibber any more than John the Baptist was demon possessed. The critics of Jesus had slandered John. They had no credibility." [305]

"But I say to you, I will not drink of this fruit of the vine from now on until that day when I drink it new with you in My Father's kingdom." -Matthew 26:29 (also Mark 14:25; Luke 22:18).

Jesus did not drink during His early ministry, and He is certainly not a drinker today. Notice this verse reveals Jesus does not drink any kind of wine today, and when He drinks with us again, it will be the new, fresh fruit of the vine.

What about the Word "Shekar"?

"And you shall spend that money for whatever your heart desires: for oxen or sheep, for wine or similar drink, for whatever your heart desires; you shall eat there before the LORD your God, and you shall rejoice, you and your household." -Deuteronomy 14:26

If all modern translations did as good a job as the NKJV on this verse, there would be little dispute. Here the Hebrew word for wine is yayin, and the Hebrew word for similar drink is shekar (spelled variously). Others translate shekar as "strong drink," "fermented drink," or even "beer."

This verse has become the "John 3:16" of Christian social drinkers. It trumps all others, and gives them license to drink any alcoholic beverage they desire; as long as they do so moderately. Curiously, this verse says nothing about moderately drinking shekar.

Shekar is one of the words from which we get our words sugar, cider, and saccharine. Other languages get similar words from shekar. Rather than always an intoxicating drink, like wine, it could also refer to a fresh, unfermented, un-intoxicating drink. It was used to signify a drink, whether alcoholic or not, made from fruit other than grapes. Shekar could be preserved in all the ways unfermented wine was preserved. While many authorities believe shekar always means an intoxicating drink, significant authorities and evidence disagree.

Why Shekar in Deuteronomy 14:26 Does Not Justify the Use of Alcohol.

1. This would be inconsistent with the rest of Scripture that both directly and indirectly condemn the recreational use of this mind altering drug.

2. It would justify the use of other hard drugs. If alcohol is condoned, why not a long list of other drugs? Especially if they are legal.

3. Why would God forbid the priests to use alcoholic wine in worship (Leviticus 10:8-9), yet then approve the use of alcohol by those worshipping?

4. There is significant evidence that shekar (variously translated strong drink, fermented drink, similar drink, and beer) had various meanings, including a sweet, un-intoxicating fruit drink.

That evidence includes works of Dr. Stephen M. Reynolds, Dr. Robert Young, Dr. Frederic Richard Lees, Dr. Lyman Abbott, Leon C. Field, Dr. Robert P. Teachout, Dr. John Kitto. The translators of the New King James version chose to translate shekar in Deuteronomy 14:26 as "similar drink," thereby pairing it with the various meanings of the word wine.

5. There is evidence of shekar referring to an unfermented drink in this verse, and in Isaiah 24:9.

Strong drink (shekar) is bitter to those who drink it. -Isaiah 24:9
Shekar would not turn bitter if it was alcoholic; through the fermentation process it would already be bitter. This statement is more easily understood with the realization that shekar could be sweet and un-intoxicating. Their sweet drink would become bitter under the judgment of God. After all, the chapter has just said (v.7), "The new wine fails, the vine languishes." The context is speaking of un-intoxicating wine. Verse 9 then speaks of sweet, new, un-intoxicating shekar becoming bitter to those who drink it. (See John Kitto on "Drink, Strong") Deuteronomy 29:6 is another likely reference to unfermented wine and shekar.

6. Shekar is usually used with yayin (wine). Yayin sometimes refers to alcoholic wine; sometimes refers to nonalcoholic wine. When yayin and shekar are paired, they signify the same thing, both referring to alcoholic drink, or both referring to nonalcoholic drink. The NKJV recognized the parallelism or hendiadys [306] (wine or similar drink). Parallelism is often used in Scripture. The context determines whether wine (yayin) and similar drink (shekar) are intoxicating or not.

7. Shekar seems to primarily be distinguished from wine in that wine, whether alcoholic or not, is made from grapes. Shekar, whether alcoholic or not, is made from fruit other than grapes, and possibly also came to be used for beer. Just as the words for wine were used in a wide generic sense, so also the word shekar.

8. Even among those who agree shekar is an alcoholic drink, they dispute whether it is a fruit drink or beer. This seems to lend additional credence to the possibility of shekar having multiple meanings.

9. No restriction is made about shekar being enjoyed by men, women, and children. No hint of moderate drinking is implied. They are to buy wine and shekar, and drink up before God. In the entire context of the Bible, and the context of the holiness of God, this is much easier to accept if wine and shekar were sweet unfermented healthy products of grapes and other fruit.

10. It is improper to take an obscure verse like Deuteronomy 14:26 and make a doctrine out of it while ignoring the plain passages condemning intoxicating drinks like Proverbs 20:1; 23:29-35; 1 Thessalonians 5:6-8 (sober). Also ignored are the biblical principles against using a recreational mind-altering drug. One of the basic rules of proper biblical interpretation is that you interpret obscure, hard-to-understand passages in the light of more frequent, easy-to-understand Scripture.

11. Shekar was a fruit drink made from fruit other than grapes. This beverage is clearly referred to in Scripture as an alcoholic drink (Proverbs 20:1) and as vinegar (Numbers 6:3). It stands to reason that it was also made into, and used as, a nonalcoholic drink. The people possessed the skill to do so. The people also had the desire for non-intoxicating sweet beverages. They had a lack of sweet things and craved them. It stands to reason that, just as they made, preserved, and desired nonalcoholic sweet drinks made from grapes, they would also do the same with apples, palm dates, pomegranates, and other fruit. If shekar existed as alcohol and as vinegar, it also existed as sweet fresh fruit juice and boiled down must. Just as you have soft cider before you can have hard cider, the same is true of both wine and shekar. You first press the fruit for the sweet, nonalcoholic juice, then work to turn it into alcohol. For alcoholic shekar to exist, it must first have been nonalcoholic, sweet shekar. They could preserve it, of course, in either the nonalcoholic or alcoholic state.

12. God is here giving His endorsement for the use of wine and shekar for recreation and worship. He gives no limit or restriction to their use. Would God give His blessing on the in-

discriminate use of hard drugs? Can you imagine a holy God giving His sanction to marijuana and cocaine in a country where they were legal?

13. The verbal form of shekar, while often referring to drunkenness, does not always do so. Sometimes it simply refers to being filled or drinking to the full. Likewise, the noun shekar is generic and refers to nonalcoholic as well as alcoholic drinks.

14. The words sugar, cider, saccharine come from the Hebrew word shekar. The idea conveyed in these other languages is that of sweetness. Date or fruit juice other than grapes would lose its sugar content through alcoholic fermentation. That sweet substances get their name from shekar would imply at least one original meaning of shekar was an unfermented, sweet drink.

Theologians, Scholars and Historians Are Not Silent

Robert Young was a linguist and produced one of the first of the modern English Bible translations. He defines **shekar** as "Sweet drink (what satiates or intoxicates)." and **sikera** as "Sweet drink, (often fermented)." [307] The Septuagint (LXX) translated / transliterated the Hebrew word shekar to the Greek word sikera. Sikera is used only once in the New Testament. William Gesenius in his lexicon adds that the verb form of sekar is "not always of drunkenness; but sometimes to drink to the full." [308]

It should be noted that the words "strong drink" are not derived from shekar and this translation has added to the confusion by giving some the impression that shekar was a strong fortified drink. They did not have the ability to distill and increase alcohol content as is often done today. They had alcoholic drinks, but not the strong, distilled drinks of modern days.

Leon C. Field details, "It [shekar] probably denoted sweet juices of all kinds originally, but came at length, in distinction from yayin, to be applied to the juices of other fruits than grapes, and, like yayin, was used generically of both fermented and unfermented drinks...As yayin is the generic term for the liquid of tirosh, so shechar is the generic term for the liquor of yitzhar or any other fruit than the grape, such as dates, pomegranates, etc...It is further confirmatory of this view that to this day the juice of the palm tree in an unfermented state, when just fresh from the tree, is a common and favorite beverage of the natives of Arabia, and is called by a name whose root is the same as that of shechar." [309]

Jerome, translator of the Latin Vulgate, c. AD 400 signifies "Shekar in the Hebrew tongue means every kind of drink which can intoxicate, whether made from grain or from the juice of apple, or when honeycombs are boiled down into a sweet and strange drink, or the fruit of palm pressed into liquor, and when water is coloured and thickened from boiled herbs." [310] Notice while his comment mentions intoxicating drink, Jerome also acknowledges it can be boiled down unfermented drinks, or a newly pressed fruit beverage. So in AD 400 the generic nature of shekar was affirmed.

Lyman Abbott was a Congregational minister, editor, and author. It was said he was not a friend to the prohibitionists, yet he agreed that yayin and shekar were not always fermented. "It is tolerably clear that the general words 'wine [yayin; oinos]' and 'strong drink [shekar]' do not necessarily imply fermented liquors, the former signifying only a production of the vine, the latter the produce of other fruits than the grape." [311]

Stephen M. Reynolds was a Presbyterian scholar with a Ph.D. from Princeton University where he worked in biblical and oriental languages. He also worked in archaeology under Dr. William F. Albright. He contributed to the NIV translation and *Baker's Dictionary of Ethics*. He indicates, "The Syriac language has a cognate word which suggests that the primitive meaning of the proto-Semitic root sh-k-r may have been a drink made from the date palm or honey (*A Compendious Syriac Dictionary* Founded Upon the Thesaurus Syriacus of R. Payne Smith, Edited by J. Payne Smith, Oxford: Clarendon Press. Article, "Shakar."). An intoxicating date wine may have, in course of time, come to be the meaning, and the word may have also taken on the meaning of beer...There is enough evidence to say that it is unjustifiable to claim that *shekar* must essentially be an intoxicating drink, and since the circumstances of its use in Deuteronomy 14:26 are such that an intoxicant is inconsistent with God's commands given in other places, we must assure that a non-intoxicant is intended here." [312]

F. R. Lees and Dawson Burns make clear, "Shakar (sometimes written *shechar* or *shekar*) signifies 'sweet drink,' expressed from fruits other than the grape, and drunk in an unfermented or fermented state. It occurs in the Old Testament twenty-three times. 'Shakar,' erroneously translated strong drink [in Deuteronomy 14:26], comes from an Oriental root for 'sweet juice,' and is the undoubted original of the European words (Greek, Latin, Teutonic, and Spanish) for sugar. It is used to this day in Arabia for palm-juice and palm-wine, whether fresh or fermented." [313]

Dr. Robert P. Teachout earned his doctoral degree from Dallas Theological Seminary, where he wrote his doctoral dissertation on *"The Use of 'Wine' in the Old Testament."* He says, "Not only the word yayin, but also shekar can refer to grape juice as well as to wine (cf. Deuteronomy 29:6; Numbers 28:7; Exodus 29:40). The Hebrew verb which is related etymologically, shakar, means 'to drink deeply' rather than 'to become drunk.' as many lexicons imply (note especially Song of Solomon 5:1). The idea of drunkenness so often associated with both the noun and the verb is dependent upon the context (and the beverage that is imbibed), then, and is not an innate meaning of the word. The two words yayin and shekar together give the one idea in Deuteronomy 14:26 of 'satisfying grape juice' (a hendiadys)." [314]

Dr. William Patton (1798-1879) studied at Middlebury College, Princeton, University of New York. He was an abolitionist and pastor in New

York City. His research provided that "Shekar is a generic term, including palm-wine and other saccharine beverages, except those prepared from the vine." [315]

John Kitto, D.D., F.S. A., was one of the most respected Bible scholars of the 1800s. He had lived in the Middle East and had an extensive knowledge of Bible times and customs. Kitto said shekar had, "In all probability a much wider signification than is now conveyed by the phrase 'strong drink.'"

Kitto then goes on to give three meanings of the word shekar. First, "Shechar, luscious, saccharine drink, or sweet syrup, especially sugar or honey of dates, or of the palm-tree; also, by accommodation, occasionally the sweet fruit itself." Second, "Date or Palm Wine in its fresh and unfermented state." Third, "Sakar…denotes both in the Hebrew and the Arabic, fermented or intoxicating palm wine. Various forms of the noun in process of time became applied to other kinds of intoxicating drink, whether made from fruit or from grain."

Kitto points out the original meaning of Shekar as a sweet drink has been carried into other languages. "Arabic, sakar; Persic and Bengali, shukkur; common Indian, jaggree or zhaggery; Moresque, sekkour; Spanish, azucar; and Portuguese, assucar." Also, "Greek, sakhar; Latin, saccharum; Italian, zucchero; German, sucher and juderig; Dutch, suiker; Russian, sachar; Danish, sukker; Swedish, socker; Welsh, siwgwr; French, sucre; and our own common words sukkar (sweetmeat), sugar, and saccharine." [316]

The Temperance Bible Commentary, demonstrate that the verb form, "Shahkar - connected as root or derivative with shakar, 'sweet drink' - strictly implies, as Gesenius states, 'to drink to the full,' generally with an implied sweetness of the article consumed, whether the sweet juice of the grape or other fruits. Whenever the juice had fermented, or had become intoxicating by drugs, this plentiful use would lead to intoxication, and give to the verb the secondary sense of inebriation in the drinker. Inebriation, however, must not be inferred unless the context suggests such a condition." [317] Two examples where it probably did not mean inebriated, but simply filled, are Genesis 43:34; and Song of Solomon 5:1.

John W. Haley has this to say about Deuteronomy 14:26, "The words rendered 'wine' and 'strong drink' may not imply here fermented or intoxicating liquors." [318]

Shekar is transliterated from Hebrew into the Greek sikera and is found in Luke 1:15 where John Baptist is not to drink wine or sikera (shekar). In his early English translation (AD 1382), John Wycliffe translated it, "He shall not drink wine nor cider." [319] The English word cider is derived from shekar. Just like wine and shekar, cider can be either nonalcoholic or alcoholic, sweet or hard. Cider is actually an example of shekar and a good English translation.

Some scholars read shekar as always referring to beer. Patrick E. Mc-Govern disagrees. "The biblical text, however, makes no mention beer, and vineyards actually provide the backdrop for the action. The Philistine 'beer-jug' might better be called a 'wine-jug.'" He goes on to say, "Nowhere does the Bible claim that sekar was a grain product, whereas making wheat and barley into bread is a constant refrain…The related Akkadian word, sikaru, provides the best clue. It is usually translated as 'date beer' or 'date wine.' With twice as much sugar as grapes, dates could be fermented to a higher alcoholic content, possibly as high as 15 % by volume. Date palms thrived in the warm, humid coastal plains and the oases of the interior. Their fruit could be concentrated down to a 'honey' or made into a powerful drink with a brownish hue, as Mesopotamia had taught the world." [320] This date wine or shekar could be concentrated down and kept unfermented, or made into an intoxicating drink.

Samuele Bacchiocchi, late Adventist professor of Church History and Theology and a graduate of the Pontifical Gregorian University in Rome, Italy, states, "Like yayin (wine), shekar is a generic term that could refer either to a sweet, unfermented beverage as suggested by Isaiah 24:9, or to a fermented intoxicating beverage as indicated in most other instances (Proverbs 20:1; 31:4-6; Isaiah 56:12)." [321]

Turning again to Moses Stuart, who taught at Andover, authored a Greek grammar, and has been called the father of exegetical studies in America, we find "Both words [yayin and shekar] are generic. The first means vinous liquor of any and every kind; the second means a corresponding liquor from dates and other fruits, or from several grains. Both of the liquors have in them the saccharine principle; and therefore they may become alcoholic. But both may be kept and used in an unfermented state; when, of course, no quantity that a man could drink of them would intoxicate him in any perceptible degree." [322]

Brad Reynolds maintains, "Moderationists misuse texts like Deuteronomy 14:26, as proof texts for their position. They take a phrase or two located in the larger context of tithing and develop a fermentology. And yet, these same individuals seem to treat Proverbs 23:29-34 (Where the context IS alcohol) as void of application. Hermeneutical [biblical interpretation] principles require us to take clear texts on a subject as authoritative to unclear texts, and to interpret the unclear texts through the filter of clear texts." [323]

With all the above stated, let us consider the other hand, assuming shekar definitely always meant an intoxicating drink. Some point out that God only tolerated it, like He did slavery, polygamy, divorce, and that His fuller revelation against alcohol was given later in Proverbs and the New Testament. Others point out that an alcoholic drink could be prepared just like killing an ox or sheep and preparing it to eat. They would not have eaten an ox raw, but properly dressed, prepared, and cooked it. Strong drink could have been prepared by simply boiling it to make it safe and or diluting it heavily with water. To those who would say that no instructions are given about preparing the wine and shekar - neither are instructions given here about preparing oxen and sheep, but they obviously knew how to properly prepare them. They would have also observed God's law about eating no blood. Would they have also been careful to avoid leaven (or ferment) in this holy celebration? This author believes wine and shekar in Deuteronomy 14:26 were nonalcoholic drinks, but the above is presented as an alternate view.

Examples of Drinks from Fruit Other Than Grapes

This evidence is numerous in ancient literature. Most anything they did with grapes, they also did with many other varieties of fruit. Dalby says Greek phoinikinos oinos was "date wine made from the semidried fruit soaked in water: a recipe given by Pliny. Date wine is also familiar from Sumerian and Akkadian literature." [324] This would be an example of what the Old Testament calls shekar, whether unfermented or fermented.

Polybus, c. 150 BC, describes Lotus Wine, "tastes like a fig or a date, but is superior to them in aroma. A wine is made of it also by steeping it in water and crushing it, sweet and pleasant to the taste, like good mead; and they drink it without mixing it with water. It will not keep, however, more than ten days, and they therefore only make it in small quantities as they want it. Vinegar also is made out of it." [325] This presents the idea of shekar.

"McGowen tells, "Granatus was and still is a very popular beverage. This is modernly and mundanely known as Grenadine. It is essentially a thick, sweet, pomegranate syrup. In period, this is primarily an Arabic beverage, but could be found in Eastern Europe by the end of our period. The Pomegranates are pressed and reduced to a syrup and kept unrefrigerated for months before use. To serve it, dilute with hot or cold water with one part sugar, or mix in a variety of other beverages." [326]

Syrup of Pomegranate – "Take a ratl of sour pomegranates and another of sweet pomegranates, and add their juice to two ratls of sugar, cook all this until it takes the consistency of syrup, and keep until needed." [327]

Mahmut Tezean notes, "Certain fruit juices have been drunk in Turkey since ancient times. Some of them are known as syrups, like rose syrup and sour cherry syrup. In the 9th century, şira (grape must, sometimes spiced) or fruit juices were known as both çakir and süçik. Grape juice was the most commonly drunk fruit juice. Apricot juice was also drunk, and was known as ubak. Today şira is drunk where viniculture is common, but it has disappeared from the large cities." [328]

Pliny calls wine that which is made from fruit such as carob, pears, apples, pomegranates, cornels, medlars, service berries, mulberries, fir-cones. [329] This list shows the wide, generic use of the word wine and provides examples of what could also be called shekar.

Herodotus, called the father of history, was a Greek historian. He says, "There are palm trees there growing all over the plain, most of them yielding fruit, from which food is made and wine and honey." [330] Xenophon, a Greek historian and soldier identifies, "Wine made from the date of the palm tree." [331] Josephus contributes, "There are in it many sorts of palm trees that are watered by it, different from each other in taste and name; the better sort of them, when they are pressed, yield an excellent kind of honey, not much inferior in sweetness to other honey." [332]

Scripture records, "fruit trees in abundance" and "first fruits of all fruit of all trees" (Nehemiah 9:25; 10:35). The Bible not only cites grapes, but pomegranates, figs, olives, apples (maybe apricots or quince), and dates. Other fruits were very likely available. While the Hebrew word shekar is not used, these are examples of the drink, and the meaning of shekar. It is a drink made from fruit other than grapes, such as palm dates or pomegranates. As shown by these references and recipes, it could be unfermented or made into an alcoholic drink.

Drink a Little Wine for Your Stomach's Sake

No longer drink only water, but use a little wine for your stomach's sake and your frequent infirmities. -1 Timothy 5:23

We find from this Scripture that the young preacher Timothy abstained from wine. As a matter of fact, he apparently did not even drink grape juice or unfermented wine. Abstainers have recognized the strict medicinal use of alcoholic wine, while they reject it as a beverage. Paul's admonition to Timothy, however, to drink a little wine (oinos) for his infirmities, may not even be an endorsement of alcohol's medicinal use. He likely was only referring to good, fresh, unfermented wine, the common drink of the day. The juice of grapes and other fruit is filled with vitamins and other good things and would supplement an otherwise deficient first century AD diet. Unfermented fruit juice has great health benefits without the harmful side effects of alcohol. The same could be said of vegetables. Eat and drink your fruits and vegetables!

With modern medication, the medicinal use of alcohol is outdated. If it must be used as a medication, beware. Just because you are sick gives you no reason to go buy a six-pack! Follow a good doctor's instructions. Beware of doctors who tend to overmedicate.

The Apostle Paul in no way endorsed drinking; at most he only permitted the limited use of alcohol as a medicine. He may not have even gone that far, and simply recommended unfermented wine for its health benefits. Peter Masters states, "At this point it must be said that we cannot be totally sure that Paul was advocating fermented wine anyway, because the Greek word oinos (the most usual NT word for wine) can sometimes mean unfermented grape juice. While in many places in the NT it undoubtedly means fermented wine, it must not be taken for granted that it always means this. The word is very plainly used in the Septuagint to cover both fermented and unfermented produce of the vine. It is a broad term, and care is therefore necessary. The fact that Timothy, Paul's close imitator, was an abstainer is certain. What he was to add to his water is not so certain, though for sake of argument we allow that it may have been fermented wine." [333] Dr. Masters is pastor of London's Metropolitan Tabernacle, the church Charles Spurgeon pastored in the late 1800s.

Ernest Gordon believes, "No better medicine for Timothy's 'stomach and oft infirmities' could have been recommended by Paul than the juice of the grape." [334]

Paige Patterson observes that "Timothy was only drinking water. Some have pointed out that unfermented wine, perhaps boiled down to a thick syrup or jelly, mixed with water, would have helped Timothy deal with the strongly alkaline water of the Middle East. Still, most Christians would agree that medicine is a valid use of alcohol, if temporarily and properly prescribed by a trusted medical doctor... The clear case of religious abstinence from wine, i.e., total abstinence, is often overlooked. Timothy is drinking only water. Then Paul said, you need the wine for medical purposes. What is to be said of the reason for Timothy's abstinence to this point?" [335] The answer is obvious; Timothy was obeying the biblical commands against drinking.

On taking intoxicating wines as medicine, Pliny said, "Moreover, how uncertain the result, whether in drinking there may be aid or poison." [336] "It requires a direct apostolic counsel to prompt Timothy to use even this wine; and the apostolic direction, as Jerome observes, has two characteristics: first, it is prescribed only as a medicine; second, he is to take only 'a little' as a medicine." [337]

Perhaps Paul was recommending Timothy use a sweet wine like the wine made at the Greek island of Lesbos. This wine was known for being sweet, mild, and un-intoxicating. Athenaeus, c. AD 220, advised, "And if any one thinks it too much trouble to live on this system, let him take sweet wine, either mixed with water or warmed, especially that which is called πρότροπος (protropos), the sweet Lesbian wine, as being very good for the stomach." [338] Notice he pointed out it was good for the stomach, Timothy's very ailment.

To further validate the superiority of fresh grape juice's medicinal properties the Mayo Clinic wrote this, "Both red wine and grape juice also contain antioxidants called flavonoids, which have been shown to increase your high-density lipoprotein (HDL, or "good") cholesterol and lower your risk of clogged arteries (atherosclerosis), and may help lower blood pressure.

Eating whole red or purple grapes has benefits, as well. Some research suggests eating whole grapes also delivers the same antioxidants that are in grape juice and wine. You also get the benefit of the fiber if you eat whole grapes." [339]

A medical doctor recommended Kenneth Stanley drink a glass of wine each day for his health. Later his doctor asked if he was doing so. "No," was his reply.

"Why not?"

"I don't drink and never have," said Stanley.

"Well," the doctor said, "in that case you can drink grape juice; it's just as good for your heart."

Exasperated, Stanley asked, "Why didn't you tell me that before?" His doctor's only explanation was that grape juice has more calories than wine. [340] Grape juice (unfermented wine) has all the benefits, without the harmful side effects and danger of alcohol.

Health Benefits from Unfermented Wine

Medical research indicates these benefits come from simple grape juice.[341]
1. Contains beneficial antioxidants.

2. Helps protect cardiovascular health.

3. Encourages flexible arteries.

4. Contributes to healthy blood pressure.

Unfermented wine, like other fruits and vegetables, is filled with vitamins. So, the Apostle Paul was right when he told Timothy to "drink a little wine for your stomach's sake and your often infirmities." Paul used the generic word, "oinos," which very likely referred to unfermented wine.

One final note on the medicinal use of unfermented wine, "September 15, 2006 (Fisher Center for Alzheimer's Research Foundation) – Older men and women who drank fruit and vegetable juices more than three times a week were 76 percent less likely to develop Alzheimer's disease than those who drank juices less than once a week, a new study shows. Researchers point to disease-fighting substances called polyphenols that are naturally found in fruits and vegetables as a possible source of protection." [342] This article did warn, however, of calories and high fructose corn syrup. So "no sugar added," or "light" grape juice and fruit drinks may be best.

Not Given to Wine; Not Given to Much Wine1 Timothy 3:3,8; Titus 1:7

Among the qualifications of pastors (also known as bishops and elders in the NT) is that they be "not given to wine." Deacons are to be "not given to much wine." So some say, maybe pastors can't drink, but deacons can drink moderately. These verses seem only to be a problem for those seeking loopholes in the biblical commandments.

To be told, "Don't drive 80 mph through a 20 mph school zone," doesn't mean it is therefore acceptable to drive 79 mph through a 20 mph school zone, or even 50 mph. To tell someone not to get drunk, is not an endorsement of drinking to the point of drunkenness, or even to drink at all.

For example, the *Temperance Bible Commentary* points out the following verses.

Do not be overly wicked -Ecclesiastes 7:17. Does this mean you can be moderately wicked?

Despise not your mother when she is old -Proverbs 23:22. Does this mean you can despise her until she reaches 55, 65? With this mentality, picture young preachers debating how old is old, and how long they can despise their mothers.

Do not rob the poor -Proverbs 22:22. Does that mean you can rob the rich, or the middle class? Will we then debate how poor is poor?

God's standard for pastors and deacons in no way endorses the use of beverage alcohol. When God says don't get drunk, that is not to be interpreted as God's saying drink all you want as long as you don't get drunk. That is an example of the Pharisees' interpretation of the commandments. Pharisees saw, "you shall not murder" as endorsing anything and everything up to physically taking a life. They saw "you shall not commit adultery" as making lust perfectly acceptable. Jesus disagreed (Matthew 5:21-28).

Even the ancients understood this, "It is becoming that ministers of the New Testament, in like manner, abstain from wine...Rulers do not drink wine...We who are rulers likewise, to the people, should not yield in the least to wine." [343] "If any one of the clergy be taken (even) eating in a tavern, let him be suspended, unless he is forced to bait [eat?] at an inn upon the road." [344]

"Not given to wine" can be translated "not near wine." The bishop, pastor, elder has already been commanded to be sober, literally "wine-less" (1 Timothy 3:2; Titus 1:8; 1 Thessalonians 5:6-8). You cannot drink and be sober. This reinforces the command to be sober by saying don't even hang out in places where the wine flows. Sometimes a preacher must be around alcohol, but he does not need to make it a practice.

"The ancient paroinos was a man accustomed to attend drinking-parties, and, as a consequence, to become intimately associated with strong drink. As the Christian bishop had been previously enjoined to be neephalion [sober], it is probably that the apostle intended by this word paroinos not so much the absence of personal insobriety, as absence from convivial entertainments where drinking was systematically practiced...[he] must not only be himself sober, but he must withhold his presence and sanction from places and associations dangerous to the sobriety of himself and others." [345]

Dr. Paige Patterson states, "The bishop (pastor) is to be free from wine (1 Timothy 3:3). One would presume that this admonition, at least in part, is for an example. If so, here again the ideal would be total abstinence for all who make up the body of Christ, i.e., the church." [346] An excellent point. Why would the pastor be urged to set a good example by not drinking, if it were perfectly acceptable for members of the church to drink? Pastors, deacons, followers of Christ are to be sober. They are to be abstainers from the drug of alcohol.

Pastors are told to be "wineless" (1 Timothy 3:2; sober, temperate), and not to be around wine (3:3). Three pastor / theologians sum it up. B. H. Carroll advised, "No man should be made the pastor of a church who drinks intoxicating liquors as a beverage." [347] John MacArthur concurs, "A man who is a drinker has no place in the ministry." [348] Paige Patterson adds, "Not only is it imperative that the preacher abstain from strong drink altogether, but in this day in which the liquor industry grips modern society with such violent tenacity, the preacher ought to take his stand firmly against the use and sale of beverage alcohol." [349]

Why Don't You Use Wine in the Lord's Supper Like They Did in the Bible?

The plain answer is they did not use fermented wine in the Lord's Supper. It would be unwise and unbiblical for a number of reasons.

1. Jesus instituted the Lord's Supper during Passover. At Passover all yeast was to be excluded. The bread was to be unleavened. Leaven is the same as yeast or ferment. Leaven was often used as a symbol for sin in the Bible. Jesus was sinless. The bread, representing Jesus' body is to be without leaven. The cup is also to be without leaven, without fermentation.

2. The Bible never even uses the word wine (oinos) when speaking of the Lord's Supper. If it did, it could mean either kind of wine. But it always uses "fruit of the vine" or "cup" to refer to the Lord's Supper.

3. The cup is symbolic of the blood of Jesus that He shed for our sins on the cross. Fermentation brings decay, alcohol, intoxicating qualities to a previously safe, pure, healthy product. Fresh, pure, unpolluted, non-poisonous, unfermented wine (grape juice) is a much truer representation of the precious blood of Jesus that purchased our salvation (1 Peter 1:18-19).

4. Some say since Passover was in the spring and grape harvest in the fall, the wine would have to have been alcoholic. That is not true. As previously noted, they practiced numerous ways of preserving wine unfermented. In ancient times unfermented wine could easily be available in the spring, and throughout the year. Even in modern times some have used communion wine and Passover wine made from raisins.

5. You are teaching children, youth, and adults that drinking is acceptable.

6. You may lead astray someone who is struggling with alcohol addiction.

7. There is no legitimate reason not to use safe, un-intoxicating, unfermented wine. Why use a dangerous drug in a solemn worship service?

8. The Old Testament (OT) priests were commanded not to use what all agree was alcoholic wine during the course of their priestly duties (Leviticus 10:8-10; Ezekiel 44:21). Pastors and believers today are priests (though not in all details like the OT priests). Why would God desire priests to use alcoholic wine in the Lord's Supper today, when He forbade them to use alcoholic wine in worship in OT times?

9. Jesus' life and ministry of fulfilling the law was symbolized by new wine (sweet, unfermented wine), and contrasted with old wine. Old, fermented wine in the Lord's Supper would directly contradict this symbolism.

Didn't the Corinthians Get Drunk at the Lord's Supper?

Some have said alcoholic wine was used at the Lord's Supper at Corinth (1 Corinthians 11) because the Bible says some of them got drunk. Therefore, it is acceptable to moderately use alcoholic wine for Communion. This is wrong for several reasons. First, if you want to prove the use of alcoholic wine for Communion, Corinth is not the best of examples. They were one of the most immature, worldly, early churches. Second, the problem was that the well-off people were bringing their own food and eating it themselves. The poorer folks had little to nothing to eat (v. 21). The word wine is not used. Third, the word "drunken" (methuo; v. 21) can be translated as intoxicated, or simply as being filled, whether with food or soft drink. It does not always mean intoxicated. Methuo is also used in John 2, and sometimes means "filled" or "satiated." The most natural understanding seems to be, "one is hungry and another is filled." Fourth, methuo is not contrasted with one who has nothing to drink, but with one who has nothing to eat. Fifth, even if they were using alcoholic wine, Paul was not praising, but condemning them.

Fruit of the Vine

Some have interpreted "fruit of the vine" as always and only fermented, alcoholic wine. Their view, however, is an interpretation, not a translation. If someone said, "fruit of the citrus tree," would orange juice be disqualified because it is not decomposed and fermented? The most natural meaning of "fruit of the vine" is the grape, and the sweet wine or grape juice that flows from the grapes. Unfermented wine or grape juice is a direct product of the vine; alcoholic wine is a man-made product. There is an old saying, "God sends us food, the devil sends us cooks." God sends us the pure, safe, fruit of the vine; man concocts and sends us the dangerous drug of alcoholic wine. My apologies to the wonderful cooks of the world!

Jews, Wine, Passover

Some attempt to prove alcoholic wine is the norm because Jews use it at Passover. There are a couple of problems with that argument. First, have all Jews ever closely followed biblical instruction? Even in Scripture they were often condemned for disobeying God's commandments. In that sense, they have been a lot like Baptists! Second, some Jews have used alcoholic wine in Passover; some have not. Elsewhere, examples have been given showing Jews practicing Passover with unfermented wine. But it has not been claimed that all Jews hold to this practice. So Jews' observance of Passover can be argued either way in connection with wine and our observance of the Lord's Supper.

The Theologians on the Corinthians, Wine, and Lord's Supper

Dr. G. W. Samson informs, "Clement of Alexandria, c. AD 200, said intoxicating wine was not present at the Corinthian abuse of the Lord's Supper because: first, women were present, and according to Greek sentiment, wine was prohibited to them; second, their eagerness in eating is the fault reproved, not their drinking; third, the contrast made is between those hungry and those filled." [350] It should be noted that Clement lived only a century after the last of the apostles.

Jim Anderson, professor at Midwestern Baptist Theological Seminary, commenting on the word Methe / Methuo states "In most cases this word would refer to the consumption of intoxicants. But it also may be used to refer to the profuse drinking of a non-intoxicant." [351]

Ernest Gordon writes, "The fruit of the vine' is not the decayed fruit of the vine. It is perhaps not realized how widely the natural juice of the vine was used in the classical world. Pliny speaks of many varieties of sweet wine, raisin wine, etc., with methods of preparation."[352] Orin Whitmore adds, "Did Jesus give to His disciples a cup in which were the 'bite of the serpent and the sting of the adder,' and tell them that cup contained that which represented His blood, His life-giving blood - shed for the remission of their sins? Do you believe it? No..." [353]

Paige Patterson's comments show great insight. "In the accounts of the Lord's Supper in the Gospels and in 1 Corinthians, the word wine (oinos) is mysteriously absent. The disciples took "the cup" and drank the "fruit of the vine." The absence of the term "oinos" is curious, to say the least." [354]

A. C. Dixon writes, "In the wine used by Christ in the Lord's Supper there was not a drop of alcohol." [355]

Deets Picket admonished, "It is grossly improper to make use of a decayed product - fermented wine - to celebrate the victory of Jesus Christ over death and decay." [356]

In fermenting wine, Stephen M. Reynolds says, "the microorganism produces what everyone agrees is a poison, that is, alcohol. It certainly is not a fitting thing that we should drink a poison when we honor Christ who died that we might have eternal life." [357]

Past SBC president, Herschel H. Hobbs was a pastor and prolific author. He writes, "The elements used in this Supper were unleavened bread and 'the fruit of the vine.' The word 'wine' is not used. Some interpret 'fruit of the vine' as wine. However, as the bread was unleavened, free of bacteria, was the cup also not grape juice? Wine is the product of the juice plus fermentation caused by bacteria. Since both elements represented the pure body and blood of Jesus, there is reason to ponder. The writer sees 'fruit of the vine' as pure grape juice untainted by fermentation." [358]

We have already seen the work of Dr. Robert Teachout. He adds, "At the original 'Lord's Supper' the clear implication of Matthew 26:29 and Mark 14:25, is that this was grape juice not wine. Notice that it is called the 'fruit of the vine' and it is prophesied that in the far future it will be imbibed 'fresh' or 'new' in the kingdom of God." [359]

Josephus' (c. AD 70) comments follow the events of Genesis 40:11 (clearly unfermented wine). Notice he describes the "fruit of the vine" and his description is clearly of nonalcoholic wine. He writes, "let him know that God bestows the fruit of the vine upon men for good; which wine is poured out to him, and is the pledge of fidelity and mutual confidence among men; and puts an end to their quarrels, takes away passion and grief out of the minds of them that use it, and makes them cheerful." [360] Notice also Josephus says this nonalcoholic wine makes men cheerful.

Full of New Wine

Others mocking said, "They are full of new wine." -Acts 2:13

The word used here for wine is the Greek word gluekos. Notice the Bible itself says these skeptics were mocking the disciples. Aristotle, Hippocrates, Plato and other ancient writers have strongly indicated new, sweet, wine did not intoxicate and was safe. Strangely, some scholars take every word of this mocking derision as truth, and conclude that gluekos or all sweet wine must have been highly intoxicating.

This ridicule is comparable to saying today a man is drunk on a soft drink or iced tea. Would some then conclude all iced tea is strongly intoxicating? Actually, this was an unintended compliment. Accusing them of being drunk on un-intoxicating wine is an acknowledgement that these disciples apparently did not drink at all. This would also fit into the biblical admonitions against drinking found in Proverbs, and the New Testament. Preachers were later commanded to stay away from wine. Apparently these disciples took those biblical commands seriously. There was no reason to accuse them of being drunk on gluekos, if they were regular drinkers of alcoholic oinos.

Some say new or sweet wine turns strongly alcoholic over time. Of course it may, if that is the intention. New wine does not turn strongly alcoholic in a year, unless it is intended to do so. And if it does turn alcoholic, it is no longer sweet. It all depends on how you preserve it and what your intentions are. But new or sweet wine could easily be kept in that un-intoxicating state. (See previous chapter on preserving un-intoxicating wine.)

Ernest Gordon teaches, "There is an indication in Acts 2:13 that the apostolic community, and that would include naturally the Lord Jesus Christ Himself, drank only alcohol-free wine. When at Pentecost the onlookers mocked the apostles they said: 'These men are full of new wine' (gleukos, 'must, sweet new wine,' Abbott-Smith). There would be no point in referring to unfermented wine as a source of intoxication and the strange actions following, if it were not generally understood that the apostles used no intoxicating wine. 'These drys are drunk on soft drink,' would be the present-day wording of the gibe." [361]

G. W. Samson reminds that Gleukos, "Has from such writers as Hippocrates and Aristotle been shown to be must, or preserved grape-juice." [362] Cyril, a church father, bishop of Alexandria, and called the "Pillar of Faith," adds, "They spoke not sincerely, but ironically." [363] The Wycliffe Bible records Acts 2:13 like this, "And others scorned, and said [saying], For these men be full of must." [364] A good translation; must is unfermented new wine.

Dr. Robert Teachout notes Acts 2:13 as "sarcasm." [365] Dr. Stephen Reynolds adds, "The word the mockers used would only have been used if they knew the Christians abstained from alcoholic beverages. The word they used, gleukos, refers to the sweetness of unfermented grape juice before the yeast has destroyed the natural sugar. Wine with alcoholic content can be sweetened by mixing in sugar before or after the microorganisms which produce the alcohol have died, but scholars do not accept the idea that such artificially sweetened wine was the meaning of gleukos. If the mockers had not known the Christians were abstainers from alcoholic beverages they would have said, these men are drunk, or words to that effect." [366]

As previously mentioned, both Plutarch, a Roman who lived in New Testament times, and Dalby, who lives in the 21st century AD, agree, gleukos was not intoxicating. "The power of alcoholic drinks to intoxicate was familiar enough, but the presence in them of alcohol and the nature of alcoholic fermentation were not understood; hence Plutarch's puzzled discussion of why gleukos, must or fresh grape juice, is not intoxicating." [367]

Old Wine in New Wineskins

And no one puts new wine into old wineskins; or else the new wine bursts the wineskins, the wine is spilled, and the wineskins are ruined. But new wine must be put into new wineskins." Mark 2:22 (also Matthew 9:17; Luke 5:37-39)

Many assume not putting new wine in old wineskins is obviously referring to wine fermenting and bursting the old wineskin. This was not the point at all. New wine that began to ferment would burst new or old wineskins. This fact was recognized by Job centuries earlier (Job 32:19). The point was to prevent fermentation. The yeast / leaven from old wineskins could infect and cause new wine, sweet wine to ferment and spoil. Cleanliness was crucial in preserving new wine. Luke mentions, "and both are preserved" (Luke 5:38). The new wine is preserved in that state of being unfermented, and the new wineskin is preserved. It should also be noted that Jesus Himself is calling "wine" (oinos) that which is unfermented.

Jerome, commenting on Matthew 9:17, says that "new skins, must be used for wine that is to be preserved as 'must,' because the remains of former ferment attaches to old skins." [368] G. W. Samson informs us of "The Roman custom of using new flasks in preparing and preserving wines permanently unfermented, lest the remains of ferment adhering to the inside of an old wine-flask should cause ferment in the corked and sealed must, is here referred to." [369] Ernest Gordon concludes that "Christ's few references to wine are easily understood. 'No man putteth new wine into old bottles.' This is because the old wineskins are full of fermenting lees. The unfermented wine must be put into clean, fresh (R.V.) skins to protect it from fermentation (Luke 5:36-38)." [370]

What about Luke 5:39?

And no one, having drunk old wine, immediately desires new; for he says, 'The old is better.' -Luke 5:39

1. This does not necessarily mean all men in existence; it is a general statement.

2. Jesus is not endorsing old wine, just saying that many do prefer old wine to new.

3. Old wine can be fermented wine or unfermented wine. Unfermented wine also ages and develops more complex flavors.

4. Whether or not this verse refers to old intoxicating wine, Jesus' point is that the new wine is better and should be preferred - just like the New Covenant offered by Jesus is better than the Old Covenant (or Testament).

Gordon sums up, "When Jesus said no one who drinks old wine desires the new, He is not commending old wine, whether or not fermented. Rather, he is condemning old wine and those who would hold to the traditions of the past instead of embracing the new covenant brought by Jesus. He is criticizing them for rejecting the new and wholesome wine of Jesus' views." [371]

What about Isaiah 25:6 where God tells us we will drink wine on the lees? That has to be strong, fermented wine, and He endorses it.

Wine on the lees could mean fermented or unfermented wine. Lees are dregs, sediments. I have pepper sauce on the lees, and have drunk tea on the lees. I love orange juice on the lees. They were all unfermented. Sometimes you pour a drink off the lees, sometimes you shake the drink and drink it with the lees. Wine is usually racked or strained off the lees. Some premium grape juice, however, has cloudiness and sediments.

Look closer at Isaiah 25:6, though, and you find the word for wine is not even used in the original Hebrew. It is a word (shemarim) that could be translated "wine," or more likely, "something preserved." In ancient times they could preserve fermented wine, and they could preserve non-alcoholic wine. Young's Literal Translation (YLT) translates Isaiah 25:6, "And made hath Jehovah of Hosts, for all the peoples in this mount, a banquet of fat things, a banquet of preserved things, fat things full of marrow, preserved things refined." Robert Young says, "Shemarim - what is preserved, sediment." [372] The *Temperance Bible Commentary* states, "Shemarim is derived from shamar, 'to preserve,' and had the general signification of things preserved. It occurs five times." [373]

In other words, God will give the best to His people. The Hebrew word shemarim could be translated wine, whether fermented or not, or it could be translated, as in YLT, as preserved things. Contrary to some moderate drinking advocates, this verse really gives no evidence either way about drinking. This verse should also be understood in the light of Matthew 26:29, Mark 14:25, and Luke 22:18, verses that strongly imply the wine will be new, pure, un-intoxicating fruit of the vine.

Proverbs 31:4-7

It is not for kings, O Lemuel, it is not for kings to drink wine, nor for princes intoxicating drink; lest they drink and forget the law, and pervert the justice of all the afflicted. Give strong drink to him who is perishing. and wine to those who are bitter of heart. Let him drink and forget his poverty, and remember his misery no more.

Note the contrast between kings commanded not to drink wine and those who are perishing, bitter of heart, in misery. The Talmud (ancient

Jewish writings) interprets this as giving drugs to those who are about to be executed. This might sanction alcohol being used medicinally in extreme cases, but it is in no way saying anyone who is feeling a little down should drink up. This is especially true when compared to clear verses and principles condemning the recreational use of drugs. Dr. Duane Garrett explains, "The queen-mother does not recommend a free beer program for the poor or justify its use as an opiate for the masses; her point is simply that the king must avoid drunkenness in order to reign properly. The comparison to the suffering poor and to their use of alcohol is meant to awaken Lemuel to the duties that go with his class and status rather than to describe some kind of permissible drunkenness" [374]

If a king needs clear thinking and sound integrity, serving the civil interests of people, does not the Christian need sober thinking and virtuous influence in serving the spiritual interests of people? Believers are called kings and priests (1 Peter 2:5,9; Revelation 1:6).

Colossians 2 and 1 Timothy 4

So let no one judge you in food or in drink, or regarding a festival or a new moon or sabbaths, -Colossians 2:16 Therefore, if you died with Christ from the basic principles of the world, why, as though living in the world, do you subject yourselves to regulations - "Do not touch, do not taste, do not handle," which all concern things which perish with the using — according to the commandments and doctrines of men? -Colossians 2:20-22

Some say this passage condemns anyone who speaks against alcohol. If so, it also condemns anyone who warns against cocaine, tobacco, or arsenic. Pro drinkers often use this passage to prove their point. It is a convincing argument only to those who do not know better. Anyone who uses this passage in favor of drinking is twisting the true meaning of the Scripture.

Colossians is condemning those who insist on following the ceremonial aspects of the law. We do not have to observe the Old Testament laws to be Christians or to be justified. But this passage does not invalidate the moral law of God. If this passage condemns abstainers, it also condemns Solomon and Proverbs 20:1 and 23:29-35. If Colossians condemns abstaining from alcohol, then Paul was condemning what he wrote in 1 Timothy and Titus saying a pastor is to be sober and not to be given to wine. It is not contradicting what Peter wrote about being sober. "Do not touch" does not refer to everything; Paul also said, "Do not touch what is unclean" (2 Corinthians 6:17). This passage must be properly read in context.

As believers we do not have to follow Old Testament dietary law. We do not have to abide by every health rule someone comes up with. But we are to follow clear biblical teaching and principles, and that includes not partaking of destructive drugs.

When taken out of context, Colossians 2 and 1 Timothy 4 may appear to bolster the argument of the drinkers. Taken in context, they do not uphold their view. This is especially true when examined in the entire context of Scripture.

"Upholders of alcoholic drinks have committed a grave error in exegesis because they have not interpreted according to the context in which these passages are found. Colossians 2:16-17 is written in opposition to Judaizers in the Church who would seek to impose Jewish dietary laws on the early Gentile Christians as well as obsolete Jewish holy days." [375] "The apostle is not alluding to a distinction of drinks as intrinsically wholesome or unwholesome, dangerous or safe, but to certain arbitrary and ceremonial fancies founded on Jewish ideas of 'clean' and 'unclean.'" [376] Dr. Donald Guthrie teaches, "Paul is here referring to any system which makes salvation dependent on the observance of certain food taboos or rigid adherence to the observance of certain days as sacred." [377]

1 Timothy 4:1-5

"Foods which God created," does not apply to alcoholic wine and other dangerous drugs. God creates the sweet, nutritious fruit of the vine. Man makes it intoxicating. There are, of course, harmful natural plants. There are mushrooms that can cause a painful death. God is not here condemning those who would say to stay away from them, any more than He is condemning those who teach abstinence from the poison of alcohol. Dr. Reynolds argues, "We know nothing of what the cultists or heretics of 1 Timothy 4:1-5 taught about beverages; but we can be sure that this passage does not make valid a claim that it should be used to condemn a Christian who teaches that Proverbs 23:31 is a command to total abstinence." [378]

Social drinkers often accuse abstainers of having no Scripture and manipulating Scripture. The fault may lie in the opposite direction.

Chapter Nine:
General Questions about the Bible, Wine, and Alcohol Use Today

Why Doesn't the Bible Specify Non-Alcoholic Drinks?

Some have said, "Why not use a Hebrew word for wine that can only mean grape juice?" Since this was not used, the charge is that obviously the other words for wine meant intoxicating wine. Hence, God is in favor of His people drinking that kind of wine.

The Bible basically speaks in the common language of the people. People don't speak in such precise terms, so why should we expect that of the Bible? For example, imagine two hard working blue-collar men in a pick-up truck. One says to the other, "Let's pull into that store and purchase a cold, refreshing, non-alcoholic beverage." Do they speak that way? Of course not. Rather the conversation would be something like, "Let's pull into the store and get a drink." That would be the likely conversation even though they were getting non-alcoholic drinks.

Commercials and bumper stickers bear the slogan, "Don't Drink and Drive." What? Don't they know there are specific words for intoxicating beverages? Will not someone see that motto and think they cannot drink grape juice and drive? Only the most obtuse, however, would make an issue of this. Everyone that is fair knows "Don't Drink and Drive" refers to the drinking of alcoholic beverages. "Drink" is used today like "wine" was used in ancient times; it can mean either alcoholic or non-alcoholic beverages. We seem to automatically know the difference. Generally when "drink" is used in a negative or prohibitive way, we naturally understand it to refer to alcohol.

You don't explain everything in detail. We learn to use shortcuts in conversation, assuming people know what we're speaking of. One of the shortcuts in conversation in Bible times was to use the word "wine" for drink, whether alcoholic or not. In Bible and ancient times, wine was used generically of both intoxicating (Proverbs 20:1) and non-intoxicating (Proverbs 3:10; Isaiah 16:10; 65:8) drinks. These generic Hebrew and Greek words for wine are the ones the biblical writers used under the inspiration of the Holy Spirit. If you don't like it, take it up with God. You could just as soon ask, why didn't Plato, Aristotle, Hippocrates, and Athenaeus use a specific word for grape juice. They used the common, easily understood language of their day. The Bible did the same. We do the same today.

One last consideration. The Bible does specify alcoholic wine in Proverbs 23; and it says not to even look at it.

Everybody Knows Prohibition Was a Total Failure.

While this book is not especially about Prohibition (1920-1933), it was not the failure the alcohol industry and media would like you to think. Lives were lost during Prohibition due to gangster activity and drug smuggling. Many more lives were saved. This was due to the reduction in alcohol resulting in fewer people developing cirrhosis of the liver and other alcohol-related diseases, fewer alcohol related accidents, fewer traffic fatalities, less crime, etc.

The true evidence of Prohibition is clear. Legalize something previously against the law, and you get more of it. Otherwise, why were the alcohol interests all in favor of repealing Prohibition? Once Prohibition was repealed, drinking began to climb. The same will happen if America begins to legalize other drugs like marijuana. Let's at least keep these other dangerous drugs prohibited. Also, let's prohibit beverage alcohol as much as possible. Dry counties inevitably have less drinking. The less alcohol is available, the fewer problems you have.

Some say we have lost the war on drugs since we still have illegal drugs in our midst. We might as well say we have lost the war on murder, since murders are still committed. A law against murder and strict enforcement will not make murders cease, but it will surely reduce the murder rate. It will save many innocent victims and give just punishment to those guilty of the crime. Since murder is still committed, that does not mean we should repeal all laws against homicide.

Another myth of Prohibition is that it brought on the Great Depression. The Great Depression was world-wide and in countries that had no Prohibition. Prohibition had nothing to do with our country's Great Depression. "The best evidence available to historians shows that consumption of beverage alcohol declined dramatically under prohibition. In the early 1920s, consumption of beverage alcohol was about thirty per cent of the pre-prohibition level. Consumption grew somewhat in the last years of prohibition, as illegal supplies of liquor increased and as a new generation of Americans disregarded the law and rejected the attitude of self-sacrifice that was part of the bedrock of the prohibition movement. Nevertheless, it was a long time after repeal before consumption rates rose to their pre-prohibition levels. In that sense, prohibition 'worked.'" [379]

William Bennett concurs, "One of the clear lessons of Prohibition is that when we had laws against alcohol there was less consumption, less alcohol-related disease, fewer drunken brawls, and a lot less drunkenness. And contrary to myth, there is no evidence that Prohibition caused any big increases in crime." [380] Bennett, a graduate of Williams College, has a doctorate in political philosophy from the University of Texas, and a law degree from Harvard. He was director of the National Drug Control Policy under President George H. W. Bush, and Secretary of Education under President Reagan.

Bennett continues, "But at least advocates of legalization [of drugs like marijuana, etc.] should admit that legalized alcohol, which is responsible for some 100,000 deaths a year, is hardly a model for drug policy. As the columnist Charles Krauthammer has pointed out, the question is not which is worse, alcohol or drugs. The question is, should we accept both legalized alcohol *and* legalized drugs? The answer is no." [381]

Other Countries Practice Moderate Drinking and They Don't Have Problems With Alcohol.

To the contrary, England, France and other countries known for much social drinking have serious problems with alcohol. For example, an article in *Time* magazine stated, "In decrying the excessive alcohol consumption of their compatriots, American and British health experts have long pointed to France with special admiration. Here, they said, was a society that masters moderate drinking. In wine-sipping France, the argument went, libation is just a small part of the broad festival of life, not the mind-altering prerequisite for a good time. The French don't wink like the English do at double-fisted drinking; they scorn people who lose control and get drunk in public. It's a neat argument. But it sounds a little Pollyannish now that France itself is grappling with widespread binge-drinking among its youth. Worse still, fully half of 17-year-olds reported having been drunk at least once during the previous month." [382]

People think everyone in those countries drink; not so. A pro-drinking 2005 report reveals 23% of Italians do not drink at all. 38% of young French women do not drink. 26% of Britons don't drink. [383]

What about Other Drugs?

Biblical principles would plainly lead us to abstain from any unnecessary use of drugs, whether legal or illegal. We are to be wise and sober. We are to deny ourselves. As Christians we are bought with a price, the blood of Jesus Christ. Our bodies are not our own. A believer's body is the temple of the Holy Spirit. We are to love God with all our minds. We are not to abuse that body or mind with destructive, unnecessary drugs of any kind. We are not to lead others astray.

One Molecule of Alcohol?

Some drinkers have an unusual argument. If wine contains one molecule of alcohol, it is therefore alcoholic wine. Hence, all wine is alcoholic. If you accept that contention, then I suppose it would be true that all wine is alcoholic. They further argue that the moment the skin of a grape is crushed the wine begins to ferment. Again, if you want to strain at details, that could be true.

Using this standard, it is also true that when a cow is killed, the process of decomposition immediately begins. So to use their standard, we would have to refer to all beef, even a fresh juicy well-prepared steak, as rotten meat.

Now let's re-enter the real world. Ariel, a nonalcoholic wine (dealcoholized wine) maker points out that for a nonalcoholic wine label to be legal, the wine must have less than ½ of 1% alcohol. They then boast their nonalcoholic wine has less alcohol than fresh orange juice. As Dr. Stein has alluded, this kind of wine would affect your bladder long before it would affect your brain.

Every fresh unfermented fruit drink will have a slight amount of alcohol. Even the human body produces a small amount of alcohol. So if you want to argue with that much detail, I suppose we are all alcoholics!

One molecule of alcohol is a ridiculous standard; even government regulations recognize such. So when this book refers to nonalcoholic wine, it is fully understood that it may have a slight amount of alcohol, like fresh orange juice. But it is a beverage that is safe and does not inebriate. It would be a beverage in the range of legally nonalcoholic wine today.

Who Cares about the Scholars of the 1800s?

Some have objected to using authorities for abstinence from the 1800s. What did they know back then? You don't have any modern authorities.

It is amusing that many who mock the temperance scholars because they lived in the 1800s, then turn around and quote their favorite authors like John Calvin from the 1500s; Charles H. Spurgeon, James Boyce, John Broadus from the 1800s; John Gill of the 1700s; a 1600s Confession; Augustine from the 300s, etc. They also seem to have no hesitation in quoting the Bible itself which dates back approximately 2,000 years; the Old Testament dates back to c. 1450 BC to its completion in c. 400 BC, the New Testament written in the first century AD.

You don't have to be living today to be a serious Bible scholar. This reveals a very short sighted view. Some of the greatest scholars of the last 2,000 years lived in the 1800s. Some of these scholars studied the Bible / Wine question in great detail, unlike most scholars today. To belittle them is a slur without substance on their ability and character. Why is a man's scholarship invalid because of the date of his birth? Why not argue the issues, rather than refusing to consider the evidence?

Another consideration has to do with the ancient preservation of food and drink. Pasteurization, electricity, vacuum packing, and freezing have revolutionized food preservation. This began in the late 1800s. Those who practiced food preservation before this time can sometimes be more helpful, than those today who get their food out of a freezer. They had a much closer connection to Bible times than we do today.

How many modern day seminary professors do you know that could tell you how to raise, kill and process cattle, sheep, goats, and chickens, especially without modern day conveniences? How many modern day scholars know how to make, process, and preserve wine without refrigeration or pasteurization, whether fermented or not? How many could grow and graft olive trees and process olive oil and amurca? When was the last time you heard of a Doctoral Seminar on ancient preservation of food and drink? How many are experts in lactic fermentation? Ancient people knew all of this and much more. On average, people of the 1800s knew much more about ancient food preservation than folks today.

This book, however, gives references for its views not only from scholars of the 1800s, but from the Bible itself. It also gives references from Classical Greek and Roman writers from c. 400 BC to c. AD 300 and from medieval times. Additionally, a number of references and books are given from the 1900s and the present 21st Century.

The charge that abstainers can only refer to outdated authorities of the 1800s is a flimsy allegation. A fact is true whether spoken in 1850 or 2011. This argument really just gives social drinkers a convenient way to ignore a mountain of scholarly evidence.

Aren't Those Opposed to Beverage Alcohol Just Legalists?

Many say legalism is believing something is wrong that is not explicitly stated in the Bible. If that is true, however, then those who oppose slavery are legalists. After all, the Bible does not actually say, "Thou shalt not own a slave." The same could be said about opposing abortion.

Some grant that you can have a quiet personal conviction, but if you say that practice is wrong for others, then you are a legalist. Following this argument, you can be personally opposed to slavery, but if you think slavery is wrong for others, you are now a card-carrying legalist. With this definition those who oppose pornographic DVDs are legalists. After all, the Bible says nothing about DVDs, DVD players, TV, Computers, or electricity. With the above definition the list of legalists would be long. The biblical "loopholes" would be vast. The "non-legalists" can then say, "Ah, the Bible doesn't exactly, specifically, precisely, in so many words, say not to do it, so go for it!"

Sure, people can get too picky, too judgmental, and demand that everyone do exactly what they say. But that's not legalism. We should be able to consider whether biblical teaching applies to a practice without hurling charges of legalism and Pharisaism. Legalism is not trying to live a godly life with biblical convictions. Legalism is not being against beverage alcohol or standing up for other biblical convictions. Legalism is a false belief that attempts to merit favor with God by the works of the law, by doing good deeds. Legalism is condemned in Romans 3:20 and Galatians 2:16. Rather than by the works of the law, we are to obtain favor with God through faith in the sacrificial death of Jesus (Romans 3:21-28); then we are to do good works.

Dr. R. L. Sumner explains, "Biblically speaking, 'legalism' is trusting in the law for salvation. In Galatians (which is a good example of legalism), the Judaizers were saying that without circumcision one could not be saved. Paul blasted that idea to smithereens! 'Legalism' is a word greatly misused and maligned by uneducated preachers and Christians today, who refer to standards about holy living as legalism. If someone, shall we say, preaches against booze (or tobacco, or movies, or dancing, or whatever) some immediately shout 'legalism,' showing their ignorance." [384] Dr. Norman L. Geisler adds, "More precisely, legalism is the false belief that keeping certain laws - whether biblical or not - can be used as a condition for meriting God's grace, whether for justification or sanctification (see Galatians 3:3). But one can legislate wise laws about human behavior without being legalistic in the biblical sense of the concept. Otherwise, laws against drunk driving and illegal immigration - and a host of other things beneficial to society - would be legalistic and, thereby, wrong." [385]

Argue your case on the merits, but don't start calling the fellow who may be winning the argument, a legalist.

Closely associated with the charge of legalism is the charge that those who oppose beverage alcohol do not believe in the inerrancy of the Bible or the sufficiency of Scripture. Frankly, it is difficult to take seriously this charge.

First, a long line of those who believe in the inerrancy of the Bible, have believed in abstinence. Many of them are quoted. Men such as B. H. Carroll, R. G. Lee, W. A. Criswell, John R. Rice, Adrian Rogers, Jerry Vines, Vance Havner, Paige Patterson, Judge Paul Pressler, Richard Land. The list is long. This is an arrogant, immature charge against giants of the faith. Second, the accusation is partly that no prohibition against alcohol is found in Scripture. That is false, as much of this book attests. Third, as stated above, it is foolish to demand that if the Bible does not precisely, word for word, condemn a vice, that vice is permissible. Biblical principles, wisdom, and common sense apply.

Dr. Richards sums up, "Some have foolishly and incredibly said if you believe in abstinence from alcohol you do not believe in the inerrancy of the Bible. As Jim Richards, and thousands of other Bible believing Christians would affirm, 'I am a biblical inerrantist. I also believe the Bible teaches total abstinence from alcohol as beverage.'" [386]

What's Wrong With Moderate Drinking?

The Bible describes fermented wine and says of that kind of wine, don't even look at it. The Bible says to be wise and sober. These commands directly conflict with moderate drinking.

Other Problems with Drinking in Moderation

* A medical, scientific study (2-25-2009) revealed women that drank in moderation increased their chances of getting several kinds of cancer. This study is very significant because it was a specific study of only those women who drank in moderation.

* Alcohol kills brain cells, causes disease, fetal alcohol syndrome, cirrhosis of the liver, crime, most of the unwanted pregnancies. The first drink of alcohol immediately affects and skews your good judgment.

* With every problem drinker, it started with the first drink.

* You may be able to handle it, but someone that looks up to you may not. You and your otherwise good living have given them an excuse to drink. In that sense, you are the worst of examples.

* Moderate drinkers contribute their money to an industry that is responsible for ruining the lives of millions. We should have nothing to do with that industry.

* Moderate drinkers and those who don't drink but say there's nothing wrong with drinking in moderation, are the best friends and advertisement for the alcohol industry.

* Moderate drinkers are more likely to have children who drink.

* Beverage alcohol is a poison; its use is abuse.

* It is unknown how many deaths are caused by moderate drinkers. The first drink dulls your judgment and reaction time. But if you have an automobile accident and are below the legal limit for being drunk, records are not kept. God only knows how many accidents have been caused by those not legally drunk, but who, because of a drink or two, still did not have full use of their faculties.

A 2011 study reveals, "The blood-alcohol content (BAC) limit in the U.S. is set at 0.08%, but levels well below this legal limit are associated with car accidents that cause incapacitating injury and death. According to the CDC, close to 30 people in the U.S. die every day in motor vehicle crashes involving an alcohol-impaired driver. This is the equivalent of one death every 48 minutes. 'Buzz kills,' says David Phillips, PhD a sociologist at University of California, San Diego. 'No amount of alcohol seems to be safe for driving.' ...Car accidents are 36.6% more severe even if alcohol was barely detectable in the driver's bloodstream, the study shows...'There is no safe level,' Phillips says." [387]

* Moderate drinking is often a gateway to heavier drinking and to other drugs.

One of the biggest fallacies of the moderation position on alcohol is the fact that beverage alcohol is a recreational, mind-altering drug. You are arguing for taking a hard drug solely for the pleasurable effect it has on you. Therefore, this opens wide the door for the "moderate use" of marijuana, cocaine, heroin, methamphetamines, and any other recreational drugs of choice.

Some moderate drinkers then reply, "But the Bible says we are to obey the laws of the land, and there is a law against marijuana and cocaine." It is true there are laws against these drugs today in America, but that may change. And there are countries today that allow them. Can an American Christian tourist visit a country where these drugs are legal and moderately enjoy them? To be consistent, the moderate drinker must also endorse the moderate use of any and all other recreational drugs. To the inconsistent moderate drinker - why is alcohol acceptable, but the other drugs unacceptable? The solution is easy. Intoxicating wine is a mocker; it bites like a snake, so stay away from it!

Someone asked the question, "What could possibly be wrong with my having a beer or two at my home on a Friday night with no one around but my family?"

My reply:

* Someone may see you buying that alcohol and be led astray. To them it simply gives permission to drink in any amount. You may be able to hold and control your liquor; they may not be able to do so.

* You would have to confess to anyone who asked, that you drink.

* You would be the best advertisement the liquor industry has. They don't want to use their best customers, alcoholics, as advertisement; they want to use someone like you.

* You would be supporting an evil industry that is our number one drug problem.

* You would run the risk of becoming addicted. Moderate drinking often leads to immoderate drinking.

* You would be teaching your wife and children to drink. You would be teaching them it is acceptable to take drugs for their own amusement.

* An emergency could arise in the evening or nighttime, and your judgment and reflexes would be lacking. Could you drive your wife or children to the hospital? Could you protect them from an intruder? Could you make a late night call to someone in need?

* Is there ever a time when a Christian should have a self-induced lack of good judgment and self-control?

* Even in moderate amounts, alcohol is destructive to the human body.

The Voice of Others on Moderate Drinking

A husband with two young daughters recently said, "If they ever call and need me to pick them up at any time of the day or night, I want to be ready to drive to them. I never want to have someone else help them because I was not fit to drive because of even moderate amounts of alcohol."

John Wesley implored, "You see the wine when it sparkles in the cup, and are going to drink it. I say, there is poison in it, and therefore, beg you to throw it away. If you add, 'It is not poison to me, though it may be to others,' then I say, 'throw it away for thy brother's sake, lest thou embolden him to drink also. Why should thy strength occasion thy weak brother to perish, for whom Christ died?" [388]

Charles Wesley Ewing preaches, "Untold millions have stumbled by believing there is no harm in moderation, but their moderation led to enslavement, and those who sanctioned moderation, by word or example, share in the responsibility for their downfall." [389] Dr. Richard Land of the Ethics and Religious Liberty Commission (ERLC) notes, "A study done a few years ago found that in homes where the parents were total abstainers from alcohol, 16 percent of the teenagers in the home experimented with alcohol before adulthood. In homes were the parents were social drinkers, 66 percent of the children experimented with alcohol before adulthood." [390]

Jack London confessed, "A poor companion without a cocktail, I became a very good companion with one. I achieved a false exhilaration, drugged myself to merriment. And the thing began so imperceptibly that I, old intimate of John Barleycorn, never dreamed wither it was leading me. I was beginning to call for music and wine; soon I should be calling for madder music and more wine." [391] London was eventually destroyed by alcohol (John Barleycorn), and died at age forty.

An unknown author proclaimed, "The great discovery has at length come forth like the light of a new day, that the moderate consumers of intoxicating drinks, are the chief agents in promoting and perpetuating drunkenness." Lieutenant General Winfield Scott had this to say, "Drinking and drunkenness among the rank and file of an army, soon become one and the same thing." [392] The "prince of expositors," Alexander MacLaren (1826-1910), reasoned, "It is very clear that if a man is a total abstainer, he can never be a drunkard. As much cannot be said of the moderate man." [393]

Nobel Prize winner Dr. Robert A. Millikan includes, "I think it is unintelligent for anyone to take into his system regularly a habit forming drug such as alcohol. The biggest social force in the world is the force of example. What we do is more important than what we say. Again, the experts tell us that driving accidents can be caused by the consumption of even one glass of beer. Even small quantities of alcohol slow down the quickness of our reactions. In half of the automobile accidents on the roads, tests show that the drivers had alcohol on the breath." [394]

Dr. Stephen M. Reynolds believes "that the USE of alcohol for non-medicinal purposes constitutes its ABUSE, just as it does for any OTHER DRUG." [395] Dr. Jerry Vines, former pastor of First Baptist Church, Jacksonville, Florida and SBC president articulates, "Moderate drinking is moderate intoxication." [396]

The Bible Says Wine Makes Glad the Heart of Man (Psalm 104:15; Judges 9:13). Now that has to be Alcoholic Wine.

That attitude reveals much about the one who says it. Does man have to have a hard drug to make his heart glad? Do you need drugs to make you rejoice?

Any farmer or gardener will tell you the joy they experience at harvest after all their long hard labor. Any good cook will tell you the joy of producing the finished culinary product of their labor. Those who partake would feel the same way. This joy would be multiplied in ancient times in an agricultural lifestyle where your life depended on harvesting and preserving food for the year.

Psalm 104:15 not only mentions wine, but oil to make his face shine, and bread to strengthen man's heart. These food products were staples of Jewish life. Their production guaranteed food and prosperity through the year. In addition, the previous verse is also speaking of food. The context is food, not drugs.

Sweet things also brought joy. Today we live in a sugar-saturated society. In Bible times cane sugar was nonexistent. Sweets were rare. Man has a craving for sweets. Imagine after a year's labor and the rarity of sweets, the enjoyment of tasting new, syrupy sweet wine that had just been pressed. The fermentation process takes away the sugar content of wine. That is one reason many preferred unfermented sweet wine. The last thing some wanted to do was to take away the sweetness.

John Haley, commenting on this passage says it, "Speaks of 'wine' which 'maketh glad' the heart of man and of 'bread' which 'strengthened' it. These two terms apparently stand, by metonymy, for food and drink. Hengstenberg says, 'What appeases hunger and thirst.' It is not an intoxicating drink which is contemplated here." [397] Josephus in the first century AD spoke of fresh pressed un-intoxicating wine and said it "Makes them cheerful." [398]

Judges 9:13 says wine "cheers both God and men." The word here for wine is tirosh which most scholars agree always, or almost always, meant un-intoxicating sweet wine. The *Jewish Encyclopedia* says tirosh, "Does not include fermented wine." [399]

There is an additional reason Judges 9:13 is actually a statement in favor of un-intoxicating wine. Notice that in this parable the vine says,

"Should I cease my new wine, which cheers both God and men." The vine speaks of "my" new wine. A vine does not possess fermented, but new, unfermented, nonalcoholic wine.

God is not in any of these passages condoning the recreational use of a hard drug. Scripture teaches a drug-free joy that involves no regret. Alcohol brings a false joy, brings regret, and magnifies the troubles you already have. This social drinking position has a crude view of not just man, but God's enjoying getting a little drunk (Judges 9:13). That is not the holy, righteous God of Scripture. God and man can experience drug-free joy.

Ecclesiastes 9:7 "Go, eat your bread with joy, and drink your wine with a merry heart; for God has already accepted your works." It has already been demonstrated that wine in the Bible and the ancient world referred to both nonalcoholic and alcoholic wine. So never jump to the conclusion that every reference to wine was to alcoholic wine. The wine of today is not the wine of ancient times. Here wine is mentioned with bread. In other words, it is simply viewed as a food, not a drug. A merry heart is mentioned. Sometimes that is connected to drunkenness, but certainly not always. I have often drunk a soft drink with a merry heart. After a hard day of physical labor, a large glass of iced tea has brought joy. Notice that bread is received with joy, just as wine with a merry heart. If this proves wine must be intoxicating to produce merriment, does the bread have to be drug laced to bring joy? Ecclesiastes 8:15 even says, "eat, drink, and be merry." Notice this verse says nothing of wine or drugs. Wine is much like the word drink; it can just as easily refer to a nonalcoholic drink, as an alcoholic drink. And this verse shows that a drug free meal can make a heart merry. See also Ecclesiastes 2:24; 3:13; 5:18 and how they parallel 9:7.

Solomon, the author of Ecclesiastes, is the same who wrote Proverbs. In Proverbs Solomon described alcoholic wine, and then condemned that kind of wine 20:1; 23:29-35). Why would Solomon condemn alcoholic wine and then praise it? The answer is obvious. He condemns alcoholic wine and condones nonalcoholic wine. It should also be recalled that nonalcoholic sweet wine was very common and sought after in biblical times. In no way does this verse endorse the consumption of alcohol.

Ewing reminds us, "The pure juice of the grape is one of God's blessings given to man for his enjoyment, benefit and health, and God's sanction is on its use. Fermented wine is a product of corruption, and it has been a corruptor of mankind through the ages." [400]

1 Chronicles 12:40 "Moreover those who were near to them, from as far away as Issachar and Zebulun and Naphtali, were bringing food on donkeys and camels, on mules and oxen—provisions of flour and cakes of figs and cakes of raisins, wine and oil and oxen and sheep abundantly, for there was joy in Israel." Notice what produced joy - flour, cakes of figs and raisins, oil, oxen, sheep - as well as wine. These products do not bring a drug-induced high, but they do bring joy to the righteous.

Proverbs 27:9 "Ointment and perfume delight the heart." Other translations use the word joy; yet ointment and perfume do not bring a drug-induced high. To argue God sanctions alcoholic wine and is pleased that it brings a drug-induced high is a weak attempt to twist Scripture to the view of those who need an excuse to indulge in recreational mind-altering drugs. God's Word in no way promotes moderate or immoderate intoxication.

1 Samuel 14:27 Jonathan "stretched out the end of the rod that was in his hand and dipped it in a honeycomb, and put his hand to his mouth; and his countenance brightened." Honeycomb is sweet, like nonalcoholic wine. A substance does not have to be alcoholic to brighten your countenance and bring joy.

I Heard that *Bible Wines* by William Patton Has Been Thoroughly Refuted

Or, you can add most any other book in favor of abstinence from alcohol. Actually, the book has only been thoroughly refuted and discredited in the eyes of those who favor drinking. Sometimes they point to a minor point in a book that is incorrect and thereby try to discredit the entire book. That is an invalid argument. Sometimes they point to a scientific or medical comment, and since that time we have gained more information showing that statement was untrue. That in no way invalidates the rest of the book. Sometimes they discredit the author. Why not debate the contentions in the book, rather than attack the character of the author? Often critics make dogmatic statements that, upon investigation, turn out to be false, or simply opinion. Sometimes an abstainer may make uneducated and invalid arguments, but there are a wealth of good, solid books in favor of abstinence. Many are listed throughout this book and in the appendix.

Saying a book against alcohol has been thoroughly discredited is just an easy way to dismiss the book without having to confront the evidence it presents. These books have not been discredited; social drinkers have just disagreed with their work. Read these books for yourself; don't just accept the critics' dismissal.

Why Would Ancient People Want to Drink Grape Juice Instead of Wine?

Think of all the nonalcoholic drinks that are enjoyed by people today. We have a wealth of un-intoxicating drinks that everyone enjoys. Examples would be soft drinks, tea, coffee, fruit drinks, milk, water, shakes and malts, sports drinks.

While ancient people did not have some of the drinks listed above, they had multiple varieties of nonalcoholic wine. Just as we have a taste for Coca Cola®, Dr. Pepper®, iced tea, hot tea, spiced tea, all kinds of coffee; they had a taste for the different variations of fruit drinks they often called wine or shekar. They mixed wine with milk. These offered a number of different strengths and flavors. And ancient wine could be strong in flavor and concentration and still be nonalcoholic.

Most today are unaware of the many different flavors available from different varieties of grapes and other fruit and their unfermented juice. If we only had these drinks, and nothing else but water and milk, we would become much more acquainted with these various drinks today.

We err when we assume they only wanted alcoholic drinks. They craved the sweetness found in unfermented wine and shekar. As in ancient times, people today do not always choose alcoholic drinks.

The Only Ones Who Believe in the Two-Wine Theory Are Those Who Are Against Alcohol in the First Place.

Some challenge abstainers to, "Just name one scholar who believes in drinking that agrees that there was nonalcoholic wine in ancient times." This argument has been made by a number of social drinkers.

First, this claim sounds suspiciously like, "Everyone agrees with our view, except for those who don't, and they don't count." Many who believe there was nonalcoholic wine in Bible times are opposed to drinking. Does that make their view invalid or cast aspersions on their character? Does their believing in abstinence negate their scholarship? On the other hand, does rejecting abstinence make you a more competent scholar? Are scholars who believe in drinking the only ones who are unbiased? This book has presented a wealth of evidence, much by prominent scholars, proving that nonalcoholic wine was commonly made, used, preserved, in the ancient world.

But back to the accusation. I do not know what position Dr. Robert Young took, but he translated Deuteronomy 14:26 as "strong drink" (shekar). On the other hand, Dr. Young recognized that shekar could be either an alcoholic or nonalcoholic drink. More than one has said Dr. Lyman Abbott disagreed with the prohibition movement of his day. Yet he agreed that yayin (wine) and shekar (similar drink or strong drink) could be either alcoholic or not. A wealth of quotes given in a previous chapter from the ancient world (Aristotle, Pliny, Columella, Athenaeus, etc.), most of them non-Christian drinkers, demonstrate they recognized there was sweet wine that for some reason did not inebriate. As far as this writer knows, Andrew Dalby and Patrick McGovern would not agree with much of this book, yet they point out nonalcoholic drinks of ancient days. Dalby especially points out the un-intoxicating wine of the ancient world. So the above accusation is easily proven false.

This argument really just gives a convenient way to cavalierly dismiss all evidence to the contrary. If their argument is accepted, no longer do drinkers have to deal with much scholarly evidence that is devastating to their view.

Prolepsis (Anticipation)?

Moderationists have devised a clever plan to disregard Bible verses that call wine what is obviously unfermented wine or grape juice (Proverbs 3:10; Isaiah 16:10; Joel 2:24). They say this is a prolepsis. In other words, sure this is grape juice called wine, but it is in anticipation of its turning into fermented wine. There are some noticeable problems with this strained argument.

1. The verses listed above are not, and do not read like prolepsis. It is obvious that the immediate unfermented product of the pressed grape is what is intended; and that is called wine. No future anticipation is indicated.

2. This is a transparent, convenient way of moderate drinkers to ignore and dismiss biblical evidence that clearly contradicts their view.

3. Even in a possible case of prolepsis, who is to know whether this just-pressed unfermented wine will be turned into alco-

holic wine or, preserved in its nonalcoholic state, turned into vinegar, made into raisins, fruit leather, preserves, boiled to a thick, nonalcoholic consistency? Why must that cluster of gapes containing wine be only used to make a fermented drug-filled product? Nonalcoholic uses of wine and grapes were more common than alcoholic uses.

4. According to this view, there would be no such thing as grape juice or unfermented wine, since it can always later be turned into fermented wine. With this reasoning, all products of the grape are really alcoholic wine. With this reasoning all meat products of cattle could be called rotten; after all, if not properly preserved, the beef will soon decompose. With this view, all grape juice or wine could be called vinegar, since that is the last stage of fermentation of wine.

5. Even if this is anticipation of this juice being turned into alcoholic wine, the fact remains that clearly nonalcoholic juice is called wine.

6. The prolepsis argument ignores the multitude of other evidence where nonalcoholic wine is called wine. Aristotle called must (new unfermented wine) a kind of wine. He spoke of the different properties of different kinds of wine, including unfermented wine. Theophrastus and others spoke of sweet and harsh or dry wine. Many spoke of the un-intoxicating characteristics of sweet wine (see *Sweet Wine* article, and *Ancient Quotes*).

Regardless of the prolepsis argument, in ancient times it was common to speak of nonalcoholic wine as wine. It was common to do this in multiple ancient languages. The prolepsis argument is a weak attempt at dismissing uncomfortable evidence and seems a case of special pleading.

Law of First Mention?

Some have argued the law of first mention. In other words, since the first mention of wine in the Bible is alcoholic wine (Noah getting drunk), therefore it follows that all wine in the Bible is alcoholic wine. This ignores clear evidence to the contrary (Proverbs 3:10; Isaiah 16:10; 65:8; see chapter on Non-Alcoholic Wine in the Bible).

The same people use the same word with different meanings in different contexts. We simply figure out the meaning of the word by how it is used. Just because you use a word the first time one way does not mean you can never use that word in a different context and with a different meaning. As previously mentioned, words such as "God/god," "Spirit/spirit," and "angel" are used in the Bible in different ways and with different meanings. The first mention of the word "God" in Scripture is of the one true God; yet later in Scripture "god" often refers to false gods. Most words in our modern-day English language have more than one meaning.

Slavery and Alcohol

Some fascinating similarities can be seen between the issues of slavery and alcohol:

1. The Bible never precisely says, "Thou shalt not own a slave." While this writer believes the Bible directly speaks against drinking alcohol, some do not. However, biblical principles certainly speak against slavery and alcohol.

2. Slavery was accepted by many Christian leaders until the 1700s and 1800s. So was alcohol. But that does not make it right. That does not mean it is invalid for believers to finally wake up to the sinfulness of either one. To say being against beverage alcohol is wrong because the church did not really make this an issue until the 1800s, could also be said of slavery. On the other hand, there have always been some within the Christian community who abstained from alcohol, even when they were the minority.

3. Some say it is permissible to personally not drink, but it is wrong to impose that on others. Would they say the same about slav-

ery? Slavery should be opposed; likewise alcohol should be opposed.

4. Articles against slavery are no more invalid because they were written in the 1800s, than those against alcohol, because they were written in the 1800s.

Some reject this correlation because they say wine is well-spoken of in Scripture, but slavery is never commended in Scripture. The Bible, however, never commends alcoholic wine; rather it is condemned.

Interestingly, in the 1800s most northern alcohol temperance organizations were also fighting against slavery. And the temperance or abstinence movement, contrary to the opinion of many, began in the northern states, not the south. Maine passed statewide prohibition in 1851.

The Lord's Supper Wine has to be Red, and Only Fermented Wine is Red.

To begin with, the Bible never says the fruit of the vine, used in Communion, has to be red. Much of the fruit of the vine used today is more purple than red. Today grape juice is almost always red to purple in color. In the early days of the church, the cup would not have been glass or clear plastic as used today. The color would not have been that evident. Alcoholic wine can be red or white. Nonalcoholic wine can also be red or white. They can also be a combination of other colors. Fresh squeezed wine or grape juice may have been more of a clear, cloudy liquid, but some grapes produce red juice as soon as they are pressed.

In the production of wine, many, but not all, grapes produce a white wine, unless the juice is left longer on the skins and lees; then it turns to red wine. Red wine, however, can be either unfermented or fermented. And the color of the fruit of the vine is not specified in Scripture. The fruit of the vine, whatever its color, can be used in Communion as a symbol of the blood of Jesus Christ, shed for the sins of humanity. If given the choice, most Evangelical Christians would probably choose a red unfermented wine or grape juice to observe the Lord's Supper. But the color is not a biblical requirement.

But Someone Told Me Martin Luther Drank.

Yes, Martin Luther drank alcohol. He was also anti-Semitic. Luther was a great Christian leader of the Reformation, but he also had some serious blind spots. Luther was wrong on drinking and wrong on anti-Semitism. Many, though not all, Christians of past centuries drank. Many of them also believed in, or ignored, slavery. That justifies neither drinking or slavery. Even Christian leaders have feet of clay and are men of their times. We can admire them for their strengths, while strongly disagreeing with some of their faults.

How Can I Promote Abstinence From Alcohol and Drugs in My Church and Family?

1. Become informed and inform others, especially teachers and leaders. Don't be unkind about it; just present the truth.

2. Preach on drugs including alcohol once a year. A good time to do so is on *Substance Abuse Prevention Sunday*, on the Southern Baptist Convention Calendar in March. Include a good bulletin insert on the subject (from ERLC, etc.). Many today have never heard a sermon against alcohol and other drugs. *Substance Abuse Prevention Sunday* gives you a good excuse to preach and teach on the subject!

3. Use the *Church Covenant* that speaks against alcohol and dangerous drugs. Put the card sized *Church Covenant* in your church bulletin, or copy it onto your bulletin, once or twice a year. Paste the *Church Covenant* in the inside cover of your hymnals or pew Bibles. Keep it in your tract rack. This can be done whether or not your church has formally adopted the Covenant. Many churches have used it since the 1850s. It can be ordered from *LifeWay* (lifeway.com.)

4. Place this book and other good books on alcohol in your church library and other public, university, and seminary libraries. Give it as a gift to church leaders, professors, parents. It can be ordered from any local bookstore, or amazon.com, or directly from the author.

5. Occasionally speak to the issue in passing in sermons and Sunday School lessons. Youth leaders can occasionally speak to the issue. Use the information and illustrations included in this book. Use illustrations from the daily news.

6. Use material and bulletin inserts from the *Ethics and Religious Liberty Commission*, and other such sources. The author of this book can be contacted for inexpensive brief articles on abstinence from alcohol.

7. Search Committees can ask a potential pastor or youth minister their views on beverage alcohol. Be aware of their views before you call them as staff members.

8. Print the 2006 SBC *Resolution on Alcohol Use in America* as a handout in class.

9. Warn your children and youth. Tell them the best policy by far is to never start or try any unnecessary drug. Give them the reasons why. Parents should never use unnecessary drugs.

10. Have a worship service, especially geared to youth, on alcohol and drugs. During the invitation ask youth who will pledge to not take a drink of alcohol, or use illegal or unnecessary drugs from this day forward, to come to the altar as a sign of their commitment. Some will not take it seriously, but some will. Use the *Pledge* found in Appendix III.

11. Speak out on the internet and other avenues concerning the negatives of beverage alcohol. Vote for and promote laws restricting alcohol. Refer people to this book and other good information on alcohol.

CHAPTER TEN:
Anecdotes and Illustrations about Alcohol, Drugs, Sin, Salvation

While I would not endorse everyone quoted, I mostly agree with these quotes. The quotes and stories have been gathered from many sources; where known, I've given credit.

Upton Sinclair and *The Cup of Fury*

Awake, awake! Stand up, O Jerusalem, you who have drunk at the hand of the LORD the cup of His fury; you have drunk the dregs of the cup of trembling, and drained it out. **-Isaiah 51:17**

Upton Sinclair (1878-1968) was a well-known, prolific author. One of his best known works was the 1906 *The Jungle*, a critique of the meat-packing industry. It was a best seller, influenced President Theodore Roosevelt, and led to the passage of the *Pure Food and Drug Act*.

Far from a Baptist preacher, Sinclair was a Pulitzer Prize winning au-thor and a liberal socialist. He knew many intellectuals whose lives were ruined by alcohol. In 1956 he authored *The Cup of Fury: A tragic, shocking record of a half-century of American genius, twisted, tortured, and destroyed by alcohol.* [401] The litany of quotes below all come from this work.

"It has been my fate to live among drinking people: novelists, poets, playwrights and stars of stage and screen. I have seen two-score of them go to their doom, eleven as suicides."

"Life is a mystery, strange beyond all telling. My inebriate father would have approved this book. When he was beginning to drink he would say, 'It is my only consolation.' But when he was getting over it, he would say liquor was a curse. I can hear him moaning: 'If only I had never touched it!' I leave that as his epitaph."

"I begin the stories of Jack London and George Sterling, O. Henry and Stephen Crane. I begin the story, essentially, of a group of brilliant and brave Americans who lived to write and died for wine."

"I once asked George Sterling about this; he would know, I felt, if any man in the world could know. He answered, 'If you write when you have been drinking, you think you have written the most wonderful thing in the world; but when you read it in the morning, you discover that it makes no sense.' Great men are great not because of alcohol, but in spite of it."

"This had been the sight I had seen all through the night - people, sodden and dazed, out for 'a good time.' And I thought how many wonderful things there are in this world, so much to do and so much to learn - and so much of it being lost in exchange for the measly, momentary 'warm glow' of whiskey."

"Thus, at times, I have nettled the feelings of friends. It has always been a difficult choice to make: is one going to lose friendships because he inveighs against drink? Or must he run the risk of losing his friends *to* drink."

"The Communists use liquor as a sort of Geiger-counter, probing for the weaknesses of men and women. They have used it to gain recruits; they have used it to steal a nation's most guarded secrets...One of the early forms their [Communists] idealism took was the prohibition of alcoholic liquor. But that hope died with all the others. The brutal men who came out on top in Russia knew that a drunken people would be easier to hold in subjection than a sober people."

"When both parents drink, 83% of the female students are drinkers, compared with a mere 23% when both parents abstain. These data suggest that parental example is a factor of major significance in drinking by young people…Yale research showed that if a student abstains and makes no special point about it, out of 10 classmates, on average 9 will either not care or admire him, only one will feel scorn."

"Not drinking is no easy passport to happiness, no automatic assurance of a good and happy life. What it does is to increase the odds enormously."

"I cast my vote against social drinking. I will not keep a dog in my house that bites one of every five or nine people who stoop to pet it. Nor will I sanction alcohol because it dooms or harms 'just' one of every five, nine, or sixteen who drink it."

"For those who have not yet had their first drink - the wisdom and courage to say 'No' is the answer."

Jack London

Jack London (1876-1916) was the loved, brilliant author of *The Call of the Wild*, *White Fang*, and *Sea Wolf*. His 1913 book, *John Barleycorn*, told his life story emphasizing his struggles with alcohol, beginning when he got drunk at the tender age of five. It is one of the most descriptive, riveting books you will read on how alcohol can affect, entice, and control a man. London even wished for Prohibition, "I can well say that I wish my forefathers had banished John Barleycorn before my time." The book has been recommended by *Alcoholics Anonymous*. (John Barleycorn is an old expression for alcohol.)

In *John Barleycorn*, [402] Jack London ponders abstinence, then rejects it. Convinced he can control his drinking, he concludes, "No, I decided; I shall take my drink on occasion…I decided coolly and deliberately that I should continue to do what I had been trained to want to do. I would drink - but oh, more skillfully, more discreetly, than ever before." Three years later Jack London was dead at the age of 40. It is disputed whether he died of suicide, morphine, or uremia. But it is certain that his death was heavily influenced by alcohol. John Barleycorn had claimed another victim. Below are excerpts of his work.

Alcohol "inhibits morality. Wrong conduct that it is impossible for one to do sober, is done quite easily when one is not sober."

"An absolute statistic of the percentage of suicides due to John Barleycorn would be appalling. In my case, healthy, normal, young, full of the joy of life, the suggestion to kill myself was unusual; but it must be taken into account that it came on the heels of a long carouse, when my nerves and brain were fearfully poisoned, and that the dramatic, romantic, side of my imagination, drink-maddened to lunacy, was delighted with the suggestion." Many who would never consider suicide in their right minds, consider it when their minds have been altered by alcohol. Under the influence, London contemplated suicide.

"John Barleycorn, by inhibiting morality, incited to crime. Everywhere I saw men doing, drunk, what they would never dream of doing sober. And this wasn't the worst of it. It was the penalty that must be paid. Crime was destructive. Saloon-mates I drank with, who were good fellows and harmless, sober, did most violent and lunatic things when they were drunk. And then the police gathered them in and they vanished from our ken. Sometimes I visited them behind the bars and said good-bye ere they journeyed across the bay to put on the felon's stripes. And time and again I heard the one explanation 'IF I HADN'T BEEN DRUNK I WOULDN'T A-DONE IT.' And sometimes, under the spell of John Barleycorn, the most frightful things were done - things that shocked even my case-hardened soul."

"John Barleycorn makes his appeal to weakness and failure, to weariness and exhaustion. He is the easy way out. And he is lying all the time. He offers false strength to the body, false elevation to the spirit, making things seem what they are not and vastly fairer than what they are."

"And is there a greater maker of madness of all sorts than John Barleycorn?"

"It is these good fellows that he gets - the fellows with the fire and the go in them, who have bigness, and warmness, and the best of the human weaknesses. And John Barleycorn puts out the fire, and soddens the agility, and, when he does not more immediately kill them or make maniacs of them, he coarsens and grossens them, twists and malforms them out of the original goodness and fineness of their natures."

Famous American author Jack London tells of his early experiences with alcohol in *John Barleycorn*. When he was a boy there was no moral influence against drinking. Everyone seemed to drink. At the age of seven London got seriously drunk from what today would be called binge drinking. He was later surprised at the lack of indignation for what he had done. Instead, people laughed and pointed him out as the one who had gotten so drunk. It seemed to him that he was admired, even heroic, for what he had done. Had there been more moral examples of abstinence, had there been moral indignation at his drinking, Jack London's life may have been saved. Instead, alcohol destroyed London's life culminating in death at the age of 40.

Dr. Lorenz and Wine

Dr. Adolph Lorenz of Vienna, Austria was a celebrated medical doctor of the turn of the last century. In 1903 at a banquet Dr. Lorenz pushed his wine glass away. He was asked if he were a teetotaler. The respected doctor replied, "Yes, I am a teetotaler. I'm not a temperance agitator, I am a surgeon. My success depends upon my brains being clear, my muscles firm, and my nerves steady. No one can take alcoholic liquor without blunting these physical powers which must be kept on edge. As a physician, I must not drink." [403]

Tidbits

"I have better use for my brain than to poison it with alcohol. To put alcohol in the human brain is like putting sand in the bearings of an engine." -**Thomas Edison** (1846-1931), American inventor, scientist, businessman.

"If liquor is to be made, it certainly ought to be produced on a creek hidden from view. We have dignified it and given it status by putting it in fancy stores on Main Street. It belongs in the gutter and the swamp and in the dark, as part of the world of evil which is its native habitat." -**Vance Havner** (1901-1986), author, great preacher.

"I have a sincere conviction that liquor is one of the chief causes of unhappiness, both to the people who drink and to those who are near and dear to them. Early in my life I decided not to touch liquor even in moderation, and I have adhered to this resolution. I am grateful for God's help during periods of stress when I might have been tempted to drink had I relied on human strength alone." -**J. C. Penney** (1875-1971), founder of the department stores that bear his name.

"Two men may play tennis or chess equally well. Give one of them a single glass of beer and he will be easily defeated by the one who abstains." -**Dr. Howard A. Kelly, MD**, Johns Hopkins University.

"What is the worst that can happen from not drinking?"

An old story is told of four young people who had been drinking and were killed in a drunken traffic accident. In a rage, one of the fathers said, "I'll kill the bartender who sold them liquor!" Going to his bar for a drink to steady his nerves, he found a note written by his now dead daughter: "Dad, we took some of your liquor. We knew you wouldn't care."

You are going to have enough problems in life on your own. Don't heap upon them the problems of alcohol.

A jailed convicted murderer said, "I've been drinking since I was sixteen years old and all the trouble I have ever been in happened when I was drunk."

Suppose you are walking alone down a dark ally. A couple of young men are coming toward you. Would you feel more secure knowing they had just come out of a Bible Study, or from drinking at a bar?

An English Puritan, Quire Bruen, was at a dinner party given by the sheriff, and a toast to the prince was proposed. Refusing to toast the prince would have been considered a high insult. As the cup of wine was passed along the line, they looked to see what the Puritan would do. He said, "You may drink to his health, and I will pray for his health," and so passed the cup.

"Whiskey? I like it! I always did, and that is the reason I never use it." -Confederate General Robert E. Lee

During the early 1900's there was a tin poster that was often seen. A beautiful young woman was featured and the caption, "Lips that touch wine will never touch mine." That is still a good philosophy for the young women of today.

A reporter interviewed a wrinkled, old man, asking him the secret of his longevity. "I drink whiskey every day, smoke all day, and dance and carouse all night." "That's amazing," said the reporter. "Just how old are you?" "32," the old man replied.

An old story is told of a woman standing near a judge. The judge was hearing the case against her husband; a case that involved alcohol. The judge said, "I'm sorry, but I must lock up your husband." "Your honor," she replied, "wouldn't it be better for me and the children if you locked up the saloon and let my husband go back to work?"

This is the debt I pay,

Just for one riotous day,

Years of regret and grief,

Sorrow without relief.

Pay it I will to the end -

Until the grave, my friend,

Gives me a true release;

Gives me the clasp of peace.

Slight was the thing I bought,

Small was the debt, I thought,

Poor was the loan at best -

God, but the interest!

-**Paul Laurence Dunbar** (AD 1872-1906), famous African American poet.

"Therefore shall they eat of the fruit of their own way, and be filled with their own devices." -**Solomon**, Proverbs 1:31

"Whoever commits sin is a slave of sin…If the Son makes you free, you shall be free indeed." **Jesus Christ,** God the Son; John 8:32, 36.

"We are sinners through and through. 'The heart is deceitful above all things, and desperately wicked; who can know it?' -Jeremiah 17:9. Wow! Why feed that thing anything that could fuel its nature. That's just what alcohol does." -**Adam**, speaking to some on the internet that were defending social drinking; 2010. (No last name given.)

A Permit to Turn Decency into Indecency

"All [liquor] licenses are a permit to turn health into disease, decency into indecency, love into estrangement, young beauty into loathsomeness, woman's modesty into coarse effrontery, mother's milk into poison, manliness into beastliness, happiness into horror, honor into disgrace, intellect into driveling idiocy, plenty into poverty, comeliness into corruption, merriment into misery…Alcohol "never touched an individual on whom it did not leave an indelible stain, never touched a home in which it did not plant the seeds of dissolution and misery, never touched a community in which it did not lower the moral tone, never touched a government in which it did not increase problems." -**R. G. Lee**, [404]

"For when the mind is impaired by wine it is like chariots which have lost their drivers." -**Isocrates**, To Demonicus, 1.32; c. 350 BC. Isocrates was a famous Greek orator.

"If it is preferred to convert these grapes into wine, then there are more processes than one, well known for 4000 years by which an utterly unintoxicating wine can be made, which is delightful to the palate, nourishing as a food, and every way preferable to fermented and intoxicating wines. By cooling processes, or by boiling, a sweet and delicious wine can be made, which will keep pure, which will be a valuable commodity. A vineyard so devoted will be lighted up by day by the radiant smiles of God and be bedewed at night with His benediction. 'Thus saith the Lord, As the new wine is found in the cluster, and one saith, Destroy it not; for a blessing is in it,' Isaiah 65:8. "That such wine can be made, that it can be preserved, and that it is wholesome, palatable and nutritious is the testimony of history. About such wine, Aristotle, Pliny, Virgil and many classical authors speak."

-B. H. Carroll, A Little More Grape, Open Letter on Domestic Wines in Texas, Waco, TX; May 11, 1887. B. H. Carroll was a leading pastor and founding president of Southwestern Baptist Theological Seminary, Fort Worth, TX.

"This 'fruit of the vine' God by natural processes makes out of the water which falls from the skies. In the clouds He brews it and in the furrows He distills it. In the chemical laboratory of nature, by slow, natural processes He ordinarily converts this water into wine. Once only, by instantaneous miracle, without fermentation or distillation, but by omnific will, 'He looked upon the water - and the conscious water saw its God and blushed.'" -B. H. Carroll, Ibid.

There Is Hope

"Jesus' blood can make the vilest sinner clean." -Hymn, *Yes, I Know* by **Anna W. Waterman**, 1920 (1 John 1:7, 9). After several years of prayer, Anna and a lady friend were able to lead Anna's alcoholic husband Charles to the Lord. He was gloriously saved and cleaned up his life. As a result, she wrote this hymn. [405]

World Health Organization on Alcohol

"Alcohol causes nearly 4 per cent of deaths worldwide, more than AIDS, tuberculosis or violence, the World Health Organization warned. Approximately 2.5 million people die each year from alcohol related causes. Alcohol is the world's leading risk factor for death among males aged 15-59, 'Six or seven years ago we didn't have strong evidence of a causal relationship between drinking and breast cancer. Now we do,' said Vladimir Poznyak, head of WHO's substance abuse unit." [406]

Descriptions of Alcohol

"Alcohol is a poison men take into the mouth to steal away the brain."
-**Shakespeare**

Strong drink is "more destructive than war, pestilence and famine." -**Gladstone**

"The devil in solution." -**Sir Wilfred Lawson**

"A cancer in human society, eating out its vitals and threatening its destruction." -**Abraham Lincoln**

"Distilled damnation." -**Robert Hall**

"An artist in human slaughter." -**Lord Chesterfield**

"The most criminal and artistic method of assassination ever invented by the bravos of any age or nation." -**Ruskin**

"Drunkenness has killed more men than all of history's wars." **General Pershing**

"My experience through life has convinced me that abstinence from spirituous liquors is the best safeguard to morals and health." **General Robert E. Lee**

"He who drinks is deliberately disqualifying himself for administration.: -**President Taft**

Sergeant York

"I used to drink liquor; drank it for 10 years; drank it until I broke the hearts of those who loved me and prayed for me. And then, one night in 1914, I knelt at the altar in a little mountain church in East Tennessee, and confessed and repented of my sins. I arose from that altar a new man in Christ Jesus, and broke with liquor forever!" [407] York was an American hero of World War I.

"The habit of using ardent spirits, by men in public office, has occasioned more injury to the public service and more trouble to me, than any other circumstance which has occurred in the internal concerns of the country during my administration. And were I to commence my administration again, with the knowledge which from experience I have acquired, the first question which I would ask with regard to every candidate for public office would be, 'Is he addicted to the use of ardent spirits?'" -**President Thomas Jefferson**

A Beer at a Baseball Game

Mrs. Alice Hatch shares the following story from her daughter Becky.

"For several years Becky had a faithful walking partner who we'll call Linda. They were members of different Baptist churches. One day as they were walking Linda asked Becky what she believed about drinking. Becky answered that she didn't drink nor did she have any interest in drinking, but that she didn't have a real problem with people who choose to drink in moderation. (Alice Hatch hastens to add that Becky's stand on alcohol at that time did not reflect the strong convictions of her parents on the matter, but is apparently typical of many young Christians today.) Then Becky's friend related the following story to explain her strong conviction that Christians shouldn't drink: Linda told of her and her husband attending a baseball game with a young man who was a new Christian. He was a new member of Linda's Sunday School class. Linda did not often drink, but on that particular night she had had a bad day at work and just wanted to unwind. She and her husband had a beer and offered to buy a beer for the new Christian. The young man drank the beer, and continued to drink several more before the night was over.

The next day Linda got a phone call from the new Christian's wife. She explained that her husband had a drinking problem and had recently accepted Jesus Christ as his Savior. Once he was saved, he mustered the courage to stop drinking. He had been sober for several months when he was offered a drink by Linda and her husband. She went on to explain that her husband had fallen off the wagon because he saw other Christians drinking and thought it would be alright for himself. The young man's wife concluded by saying, 'I thought my husband was safe with Christians.'

Linda was devastated. She and her husband determined never to drink again. She told Becky that she now understood the principle of the weaker brother found in Romans 14:21.

Becky changed her convictions on drinking and often uses this story to explain the passage in Romans and how it so clearly applies to alcohol." -from **Mrs. Alice Hatch**, wife of John A. Hatch, pastor of First Baptist Church, Lake Jackson, Texas; 2009. From personal correspondence.

The General

At a banquet it was noticed that a general turned down the wine glass at his plate. A lady asked him, "Do you ever drink wine, General?" "No madam, never," was his courteous reply.

"I don't want to be impertinent," said the lady, "but I'd like to know why a person of your age and character shouldn't enjoy the pleasure of an occasional glass of wine." "Perhaps an occasional glass of wine wouldn't hurt me," said the general, smiling. "But that young fellow over there" - he pointed out a young man at another table - "is my son. If I don't drink, he won't. If I do, the chances are he'll follow my example. I turn down the wine glass, and you see he has done the same." -Knight

Meet God Sober

A physician of the 1800s advised an ex-alcoholic to take brandy for his health. The patient refused and included that he had signed the pledge. The doctor said if you don't, you may be a dead man. The recovering alcoholic calmly replied, "Well, then, if it be so, I'll meet my God sober. I'll take no more brandy." -G. W. Bengay

Billy Sunday

"Goodbye, John. You were God's worst enemy. You were Hell's best friend." -famous **Evangelist Billy Sunday** in a sermon at the beginning of Prohibition, Norfolk, Virginia, July 16, 1920. "John" referred to "John Barleycorn," an old term used in the 1800s and early 1900s to refer to beverage alcohol.

"I've been a prohibitionist ever since I got religion, and if you're not, you need another dip." -**Sam Jones**, Methodist Evangelist; quoted by Coker.

"O thou invisible spirit of wine, if thou hast no name to be known by, let us call thee devil!" -**William Shakespeare**, *Othello*.

Prohibition For Pilots

"Recently the whole country was upset that a couple of airline pilots reported for duty drunk. The country at least supports prohibition when it comes to airline pilots." -**Dr. R. L. Sumner,** editor *The Biblical Evangelist*.

No Practice So Foul

"There is no practice so foul that liquor does not gladly aid. Liquor deadens all sense of spiritual values." -**Dr. John L. Hill**, quoted by Hearn.

Resisting Temptation

"Indulgence in alcohol, even in very small quantities, weakens the power of resisting temptations." -**Dr. Emil Kraepelin**, quoted by Hearn.

Breaking Down Moral Inhibitions

"The effect of a little wine or beer upon an adolescent girl in breaking down her normal social and moral inhibitions is notorious. The effect is produced by premeditation of companions of both sexes who desire to lower the intended victim's level of behavior." -**Dr. William Healy**, *The Individual Delinquent*, quoted by Hearn.

Of Sponges and Kings

When Philip, king of Macedon, was praised as a jovial man who would drink freely, Demosthenes replied, "That is a good quality in a sponge, but not in a king." -Bengay

A Ruinous Example

"I leave the world a wasted character and a ruinous example; I leave to my parents so great a sorrow as in their weakness they could possibly bear; I leave to my brothers and sisters so much shame and dishonor as I could have brought to them; I leave to my wife a broken heart and a life full of shame; I leave to each of my children poverty, ignorance, a bad character, and the memory of their father lying in a drunkard's grave and having gone to 'a drunkard's hell.'" -found on a chair in the room where an alcoholic had committed suicide. Quoted from early 1900's literature.

Evangeline Booth on Drink

"Drink has drained more blood; hung more crepe; sold more homes; armed more villains; slain more children; snapped more wedding rings; dethroned more reason; wrecked more manhood; dishonored more womanhood; broken more hearts; blasted more lives; driven more to suicide; and dug more graves - than any other poisoned scourge that ever swept its death-dealing waves across the world." -**Evangeline Booth**, wife of the founder of the *Salvation Army*.

"One reason I don't drink is that I want to know when I am having a good time." -**Lady Astor**

"Wine is a turncoat; first a friend and then an enemy." -**Henry Fielding** (1707-1754), English novelist and dramatist.

"If you wish to keep your affairs secret, drink no wine."-**Author Unknown**

Alcohol Takes the Best

"If we take habitual drunkards as a class, their heads and their hearts will bear an advantageous comparison with those of any other class. There seems ever to have been a proneness in the brilliant and warm-blooded to fall in to this vice. The demon of intemperance ever seems to have delighted in sucking the blood of genius and generosity." -**Abraham Lincoln**, address to the Washington Temperance Society, Springfield, Illinois, 22 February, 1842.

Weakened by Alcohol

"A person weakened by the drinking of alcohol will take a cold or contract pneumonia or some other disease more readily than a total abstainer would and will take longer to recover from it." -Dr. C. Aubrey Hearn, graduate of Howard College, Vanderbilt University, received his doctorate from Atlanta Law School, and attended Yale School of Alcohol Studies.

Doctors Carelessly Prescribing Alcohol

"A physician who specialized in the treatment of alcoholics deplored the unscientific and careless habit of doctors of prescribing alcohol for many common ills. 'Repeatedly,' he says, 'patients lay the blame for their problem-drinking at the door of physicians. I recall a score of women who claimed their drunkenness was due to their doctors who counseled them to take a hot whiskey for any of a dozen minor ills…'" -**Dr. C. Aubrey Hearn**

"Its [alcohol] worth as a medicine is practically nonexistent." -**Robert S. Carroll, M.D.**, quoted by Hearn. Contrary to popular opinion, alcohol does not warm someone that is cold. It is only an illusion, much like other claims of alcohol. Actually, alcohol makes the skin temperature rise, while the body heat drops below normal. -from Hearn

"Under a tattered cloak you will generally find a good drinker."
-**Spanish Proverb**

"Wine hath drowned more men than the sea." -**Thomas Fuller**

"Alcoholism isn't a spectator sport. Eventually the whole family gets to play." -**Joyce Rebeta-Burditt**

"Wine in, wit out."

Better to Not Start, Than Start and Then Stop

When it comes to drugs, including alcohol - it is much easier to not start, than to start and then stop. Once started on drugs, you will have to deal with demons you never had to deal with before. Some are never able to stop. Danger, stay away, just say no.

A teacher asked a boy, "What is syntax?" The student replied, "The duty on liquor."

"Surely the time has come for a careful, persistent, and persuasive presentation of the fact that [alcohol] abstinence makes sense." -**Christianity Today**, 1964.

Guarantee Against Alcoholism

"I can give you a 100% guarantee that you will never become an alcoholic. Just never take that first drink." -**Joe Brumbelow**, Texas Baptist preacher.

Priceless Possession of Youth

"When a young life starts out from home to fight the battles which must be fought, one of the most priceless possessions, one of the greatest safeguards he or she can have is that of total abstinence from all alcoholic liquor." -Distinguished London surgeon, **Dr. Arthur Evans**, quoted by Hearn.

Influence

After speaking of the danger of alcoholism, "I abstain for another unquestionable reason, namely, I desire to avoid completely any responsibility for causing by example, my own children, or the sons and daughters of other parents, to become victims of alcoholism and all the other tragedies which result from alcoholic beverages." -**Dr. Andrew C. Ivy**, scientist; quoted by Hearn.

"The abstainer never has to worry about becoming an alcoholic. Total abstinence from all alcoholic beverages is the cheapest, the healthiest, the most scientific, and the wisest course for everybody." -**Dr. C. Aubrey Hearn**

Beau Rosser, pastor of Second Baptist Church, Highlands, TX, before becoming a pastor, was at times a moderate, at times a heavy drinker. October 14, 2010, he gave his testimony at the Baptist Student Ministry of Lee College. It was a powerful, outstanding testimony. Some of his statements: "For four years I never missed a day without drinking. After that I continued drinking, just not necessarily every day...During a 13 year period of time, I estimate I spent $86,000 on alcohol and other drugs...I quit drinking for a while and decided I had it licked. Now, I would be able to drink moderately. My wife and I went to a restaurant and I decided it would be safe for me to drink a beer. When the waitress brought the bill I was outraged. I called her over and said there was no way I drank six beers. By the time I was finished, she walked away crying. My wife said, 'Honey, I love you, but you drank six beers.' It was then I realized I had to quit drinking altogether...I've not had a drink in 16 years. Today I'm 41 years old. I take two insulin shots a day. My liver does not function properly. I take 14 pills every day. God forgave me, but I'm paying the ticket...You don't have to drink. It's a choice. If you don't start, you never have to stop." [408]

Turning Beer into Furniture

A skeptic once spoke up in a crowd, "What I want to know is how Jesus turned water into wine. I don't believe it." After some silence a man stood up. "I don't know how Jesus turned water into wine. It was a miracle. But I can tell you this. I was a drunk. I spent all my money, all my family's money on beer. I lost my job, lost my furniture, my wife and kids went hungry. We lost everything. Then, I met Jesus as my Lord and Savior. He changed me and He took away that beer and turned it into furniture, a nice home, and a wife and kids that are not afraid of their daddy. I don't know about turning water into wine. But I've sure seen Jesus turn beer into furniture."

Diogenes

At a feast Diogenes was given a goblet of wine. He threw it on the ground. When reproved for wasting so much good wine, he answered, "Had I drunk it there would have been a double waste, - I as well as the wine would have been lost." -Bengay

"The more drams, the fewer scruples."

Jerome

"I would begin by urging you and warning you as Christ's spouse to avoid wine as you would avoid poison. For wine is the first weapon used by demons against the young. Greed does not shake, nor pride puff up, nor ambition infatuate so much as this. Other vices we easily escape, but this enemy is shut up within us, and wherever we go we carry him with us. Wine and youth between them kindle the fire of sensual pleasure. Why do we throw oil on the flame — why do we add fresh fuel to a miserable body which is already ablaze." [409]

The Flagpole

Years ago in western New York, on election morning, the town drunk asked for a ballot to vote to prohibit the sale of alcohol. The liquor seller thought it to be a joke. He watched as the man who had struggled for years with alcoholism cast his vote for prohibition. They began to laugh at him. One said, "A great temperance voter you are. Why, if there was a bottle of whisky yonder on top of the flagpole, you'd risk your life and climb the flagpole to get it!" The alcoholic replied, "Yes, you're right. But another thing is true. If the whisky wasn't there I wouldn't climb that flagpole." -**Walter B. Knight**

Worst Bus Accident - Caused by Drunk Driver

On May 14, 1988, a repeat DUI offender with a .24 percent blood alcohol concentration, heading the wrong way down the highway in a pickup truck, slammed into a church bus returning from a trip to Kings Island Amusement Park. The fiery crash killed 27 passengers, 24 youth and three adults. Thirty others were injured. At that time in Kentucky, the level considered to be "intoxicated driving" was a blood alcohol level of .10 percent. The bus was from First Assembly of God, Radcliff, Kentucky. The accident occurred on Interstate 71 in Carroll County, Kentucky. The drunk driver, Larry Mahoney, spent almost 11 years in jail for the crime. Mahoney claimed to have no memory of the accident. A mother who lost a child in this accident, later became president of MADD. -Some information from **MADD** (Mothers Against Drunk Driving).

Educated Do Not Believe in Abstinence?

Many have argued that abstinence has only been advocated by those who are ignorant and uneducated. The many scholars quoted throughout this study prove otherwise. An additional quote also drives home this point. During the years leading up to Prohibition, "University presidents, seminary professors, medical professionals, linguists, Classics scholars, New Testament and Old Testament theologians and scholars alike argued tirelessly in professional journals, books, pamphlets, and speeches not only the personal virtue of abstaining from intoxicating beverages but also the public vice of manufacturing, distributing, selling, and consuming alcoholic beverages for social and recreational purposes." -**Dr. Peter Lumpkins**, *Alcohol: Abstinence in an Age of Indulgence*, Hannibal Books; 2009.

Nancy Reagan, Just Say No

One of the best campaigns ever against drugs was led by First Lady Nancy Reagan. The theme was simple, "Just Say No!" Others have put it, "Abstinence, it works every time it's tried."

Parents, churches, government should continue this theme.

Church Covenant

"To abstain from the sale of, and use of, destructive drugs or intoxicating drinks as a beverage." -a portion of the **Church Covenant**, Broadman Church Supplies, Nashville, TN (lifeway.com). Used by many churches since the 1850s.

The Cost of Sin

"Sin will take you farther than you want to go, keep you longer than you want to stay, cost you more than you want to pay."

A Can of Beer

"I've never seen anything good come out of a can of beer." -**Lendell Martin**, Bass Fishing Pro; quoted in Baptist Press (bpnews.net; texanonline.net), story by Jerry Pierce, 3-14-2003.

"Even if the Bible said nothing about alcohol, common sense would tell us to have nothing to do with it." -**Mark Brumbelow**, pastor, Grace Baptist Church of Wild Peach, Brazoria, Texas, 2009.

"The Christian who is fully dedicated to making a better world will scrupulously avoid 'the befuddling beverage,' for no drinker can live at his best. No one can honor God by drinking. Total abstinence is an aid to Christian growth and personality development." -**Dr. C. Aubrey Hearn**

Alcohol is a Drug

Alcohol is a drug. Just like marijuana, cocaine, heroin. When a man drinks, he is taking drugs. A man in the alcohol industry is a drug pusher. Don't sugar coat it, don't excuse it. If you have anything to do with beverage alcohol, you are involved in drugs. Men do not drink alcohol for the taste. If it is solely for the taste, non-alcoholic drinks can be mixed that taste just like alcohol. Men drink because of the drug effect; that and nothing less.

Drinking Only for the Drug Effect

"Remember that alcohol is used solely for its effect (or for some degree of inebriation), never to satisfy thirst or to cool off, since it does the opposite." -**William B. Terhune, M.D.** [410] This statement is made, not by an abstainer, but by one who advocates the moderate use of alcohol.

The Two Wine Theory has been Thoroughly Disproved?

The view that wine in the Bible and in ancient times referred to unintoxicating wine, as well as intoxicating wine, is sometimes called the "two wine theory." Some authoritatively say it has been thoroughly disproved. This view has not been disproved. It has been disagreed with, but never disproved. A wealth of concrete, scholarly evidence has been presented in favor of this view. Ancient documentation and biblical evidence have shown the truth of this view. Even if it were disproved, however, there would still be more than ample reasons to oppose drinking. Reasons such as biblical principles, common sense, danger of addiction, crime, disease, poverty, destruction of homes, auto wrecks caused by alcohol.

Cooking Wine Down the Drain

A veteran of World War II, when James E. Selkirk married Marie, he was not a Christian. After some time, Jim made a public profession of faith in Jesus Christ, was baptized, and joined the church. Unbeknown to Jim, Marie was still skeptical. Until he overheard Marie talking on the phone to one of her lady friends. Marie told her friend she was worried that Jim had just feigned his salvation to make her happy or to quiet her down. She was not sure he was sincere, she told her friend, until she came home one day and found Jim had poured all her cooking wine down the drain. After overhearing that conversation, for years Jim enjoyed telling that story. -from **James E. (Jim) Selkirk**, Deacon at Northside Baptist Church, Highlands, TX; 2009.

Abraham Lincoln's Mother's Advice.

"Men become drunkards because they begin to drink; if you never begin to drink, you will never become a drunkard." -**Mrs. Nancy Lincoln's** advice to her young son, Abraham Lincoln. Abe took her advice, to his lifelong benefit. -William M. Thayer.

Three Foes of Britain

"We have three foes - Germany, Austria and drink - and the greatest of these is drink." -**David Lloyd George**, British Prime Minister; 1917. Quoted by John Kobler.

Baseball Star Killed by Drunk Driver

Nick Adenhart was a rookie pitcher for the Los Angeles Angels. The major leagues had been a dream come true. In his fourth Major League game, he pitched six scoreless innings against Oakland, his best game. Just hours after the game, Adenhart and two friends were hit and killed by a drunk driver who ran a red light. Andrew Thomas Gallo, 22, was charged with three murder counts, drunken driving, and fleeing the scene of an accident. -from news reports, April 10, 2009.

Drink-A-Day Guideline Not for Women

A scientific study of 1.3 million middle-aged moderate drinking British women found that one glass of alcohol significantly increased their risk of contracting several types of cancer. It is noteworthy that this was not a study of problem drinkers, but of moderate drinkers. "If you are regularly drinking even one drink per day, that is increasing your risk for cancer," said Naomi Allen of Oxford University, who led this study. -from **Houston Chronicle** article, 2-25- 2009. (see also, March 4, 2009 *Journal of the National Cancer Institute*).

Women, Alcohol, Cancer

A study was done by the Cancer Epidemiology Unit at the University of Oxford and published in the Journal of the National Cancer Institute. It was funded by Cancer Research UK, the UK Medical Research Council, and the NHS breast screening program. In a study of over one million women, researchers estimated that alcohol accounts for 11 percent of all breast cancers in the UK. That means that every year, 5,000 women get breast cancer who wouldn't have gotten it without drinking alcohol. **Journal of the National Cancer Institute.** [411]

"Alcohol has many defenders, but no defense." -**Abraham Lincoln**, American president and lawyer. It seems he used this statement for both alcohol and slavery. Both are true.

"Every alcoholic began as a moderate drinker."

Five Children Killed in Drunk Driving Accident

Chanton B. Jenkins, 32, drove his car into a rain swollen drainage ditch April 18, 2009. He survived; five children, including three of his own, drowned. Dreton Travon Thompson, 11; his brother, Malik Barlow, 7; Devin Deshawn Jenkins, 4; Hallie Briann Jenkins, 4; Karrinton Jenkins, 1. A breath test 2 ½ hours later showed Chanton Jenkins was legally intoxicated. He was eventually charged with five counts of intoxication manslaughter. -from **Houston Chronicle** reports, 4-22-2009. There is no end to these stories of devastation caused by beverage alcohol.

What would Jesus do?

"So, what would Jesus do when it comes to contemporary alcoholic beverage consumption? Probably he would do what he did. And that is to utilize only beverages that have absolutely zero chance of causing inebriation. In our modern context, in my opinion, where healthy non-alcoholic drinks and water are readily available, it would be very probable that Jesus would be a total abstainer." -**Dr. R. Philip Roberts**, president. Midwestern Baptist Theological Seminary, 7-20-2006, BP.

The Saloon

A bar to heaven, a door to hell,

Whoever named it, named it well.

A bar to manliness and wealth,

A door to want and broken health,

A bar to honor, pride and fame;

A door to grief and sin and shame,

A bar to hope, a bar to prayer;

A door to darkness and despair;

A bar to honored, useful life;

A door to brawling, senseless strife.

A bar to all that's true and brave;

A door to every drunkard's grave.

A bar to joys that home imparts,

A door to tears and aching hearts.

A bar to heaven, a door to hell,

Whoever named it, named it well.

-written by a life-time prisoner in Joliet Prison

Change Your Hitching Post

In the days before the automobile, a certain cowboy was widely known for his heavy drinking. He became a Christian and resolved to abstain from alcohol. But on his visits to town he continued to tie his horse to the hitching post in front of the Saloon. An older man in the church noticed this and said, "George, I've been a Christian a long time and hope you will pardon me if I make a suggestion. No matter how strong you think you are; you need to change your hitching post."

"Even if I were not a Christian I would have nothing to do with alcohol. There is simply too much sorrow and heartache connected to it. Avoiding this devastating drug is simply the wise thing to do." **-Dr. Danny Akin**, president, Southeastern Baptist Theological Seminary; BP article, 2006.

If Alcoholism Is A Disease

1. If it is a disease, it is the only disease that is contracted by an act of the will.

2. It is the only disease that requires a license to propagate.

3. It is the only disease that is bottled and sold.

4. It is the only disease that requires outlets to spread it.

5. It is the only disease that produces a revenue for the government.

6. It is the only disease that provokes crime.

7. It is the only disease that is habit forming.

8. It is the only disease that is spread by advertising.

9. It is the only disease for which we are fined for contracting.

10. It is the only disease which brings death on the highways.

11. It is the only disease without a germ or virus cause, and for which there is no human corrective medicine.

12. It is the only disease that bars the patient from Heaven, for no drunkard (unless he repents and trusts Christ as his Savior) shall inherit the kingdom of God (1 Corinthians 6:9-10). -unknown, from old temperance literature. Or, could it be that alcoholism has the characteristics of both sin, and a disease?

"Doctors tell us that the necessity for drink becomes a physical disease. Yes; but it is a disease manufactured by the patient, and he is responsible for getting himself into such a state." -Alexander MacLaren in Portrait of a Drunkard.

Keeping Your Wheels on the Pavement

There's an old story about a preacher who was asked to drink and do other questionable activity by the fellow to whom he was witnessing. "You need to get down on my level if you really want me to listen." The preacher always graciously refused. One day the preacher found his friend with his vehicle stuck in the mud in the ditch. The preacher's offer of help was gladly received. The preacher said, "First, let me back my pickup into the ditch." The fellow cried, "No preacher, if you do that we'll both be stuck in the ditch." The preacher replied, "You're right, if I'm going to be able to help you, I've got to keep my wheels on the pavement. That's what I've been trying to tell you all along."

Get $1; Spend $8.95

"For each dollar in alcohol and tobacco taxes and liquor store revenues that federal and state governments collect, the report said they spend $8.95 dealing with the consequences of substance abuse and addiction." -**Baptist Press**, bpnews.net; 6-10-2009.

Old Rye Makes A Speech

I was made to be eaten,

And not to be drank;

To be thrashed in a barn,

Not soaked in a tank,

I come as a blessing

When put through a mill;

As a blight and a curse

When run through a still.

Make me up into loaves,

And your children are fed;

But if into a drink,

I will starve them instead.

In bread I'm a servant,

The eater shall rule;

In drink I am master,

The drinker a fool.

Then remember the warning,

My strength I'll employ:

If eaten, to strengthen;

If drank, to destroy.

-Edward Carswell

"When the Bible appeals for persons not to be filled with wine, but be 'filled with the Spirit' (Ephesians 5:18) it is teaching persons to chose the true God, the Holy Spirit, not Baccah [god of wine]. It means let the Holy Spirit control your body." **-Dr. Nelson L. Price,**

Definitions:

AD - from the Latin, *Anno Domini*, meaning, "In the year of our Lord."Does not mean after dead. AD should come before the date, BC after the date (AD 279; 950 BC).

Amphorae - a clay pot or flask used in ancient times to store wine or other products.

Bacchus - Greek/Roman god of wine.Also called Dionysus and Father Liber. Because of its association with drunkenness and immorality, the worship of Bacchus was banned for a time in the 100s BC.

BC - Before Christ.Before the birth of Jesus Christ in Bethlehem. Jesus, God the Son, has always existed (John 1), but was incarnated, made flesh, when He was born of the Virgin Mary. Actually, the calendar is a little off, and Jesus was born approximately 5 BC.

Circa - about, approximately. Often used with a general date and abbreviated, as in c. AD 320.

Dram - old term for a glass or measure of alcohol.

Grafting - taking a scion or cutting from one tree, and attaching (grafting) it to a different tree. The growth from the scion up is genetically different from the rootstock. Grafting was well known in biblical times (Romans 11) and in the ancient world. Grapevines, olives, and fruit trees were often grafted. Grafting was well known by at least 300 BC.

Hogshead - a large barrel holding from 63 to 140 gallons. Measurement especially of liquids. Often used of alcoholic beverages. Term used after Classical times. Also may refer to smaller sized barrels.

Inspissated - thickened, dried, or made less fluid by evaporation. Old literature sometimes uses this term for boiled or concentrated wine.

John Barleycorn, Ardent Spirits, Pernicious Beverage, Demon Rum - all old names for beverage alcohol.

Neat Wine - undiluted wine.

Shekar - juice of fruit other than grapes, whether unfermented, fermented, or vinegar. Word from which other languages get their words for sugar, saccharine, cider. Shekar also came to be used as a word of beer. Translated variously as strong drink, similar drink (to wine), beer. Verbal form of shekar means to be filled, satiated, or drunk.

Wet & Dry - in the early to middle 1900s it was common to refer to those against liquor as "dry," those for alcohol as "wet." Hence the wet forces, and the dry forces.

Wine - product of the grape whether un-intoxicating or intoxicating. Used in ancient times of grapes on the vine, new pressed juice, unfermented wine, fermented wine, vinegar. Sometimes just used of alcoholic wine, sometimes not.

Cyclops

In Greek mythology the Cyclops was driven mad by his first taste of wine. He was not the last.

Grief Banished By Wine

Grief banished by wine will come again,

And come with a deeper shade,

Leaving, perchance on the soul a stain,

Which sorrow hath never made.

Then fill not the tempting glass for me;

If mournful, I will not be mad;

Better sad, because we are sinful,

Than sinful because we are sad.

-**Sir W. a'Beckett**, *Chief Justice of Victoria; from Temperance Bible Commentary*

Leviticus 10:9

Some take such passages as an endorsement to drink moderately and that priests were free to drink when not on duty. This is an argument from silence. It equates to telling a teenager not to drink at a party, so he interprets it as okay to drink anywhere else, just not at that particular party. Don't argue and look for loopholes to find how much a Christian can get away with. Scripture here is plainly speaking against, not for drinking.

Presidential Declaration against Ardent Spirits

Being satisfied from observation and experience, as well as from medical testimony, that Ardent Spirit, as a drink, is not only needless, but hurtful, and that the entire disuse of it would tend to promote the health, the virtue, and the happiness of the community, we hereby express our conviction, that should the citizens of the United States, and especially the young men, discontinue entirely the use of it, they would not only promote their own personal benefit, but the good of our country and the world. Signed by: *James Madison, Andrew Jackson, John Quincy Adams, M. Van Buren, Franklin Pierce, Abraham Lincoln, John Tyler, Z. Taylor, Millard Fillmore, James K. Polk, Hanes Buchanan, Andrew Johnson*

This *Presidential Declaration against Ardent Spirits* was presented to presidents over a number of years by Edward C. Delavan. Above statement is found in *Temperance Essays* (fourth edition), collected and edited by Edward C. Delavan, National Temperance Society, 172 William Street, New York; 1869. Delavan looked upon all intoxicating drinks as Ardent Spirits.

Boiled Wine

"In the Nov/Dec 2004 issue a letter from B. Galioto intrigued me. The 'boiled wine' spoken of in the letter was probably like our New England 'boiled cider' or cider syrup, and is probably done the same way. Different liquids need different boiling times. Maple syrup is boiled for a very long time and the volume reduction ratio is 40:1. (40 gallons of sap are boiled down to one gallon of syrup.) Most liquids are only in need of being reduced by boiling to a syrupy consistency." -**Countryside Magazine**

Roman Boiled Wine

"Boiled wine is common in Roman recipes; it is, pretty much as it sounds, wine which has been reduced to a syrup by boiling." -**John Edwards,** *The Roman Cookery of Apicius*, translated and adapted for the modern kitchen; 1984. 3owls.org

Ancient Abstainer

King Burebistas of Dacia was strongly against drinking, even to ordering the destruction of all vineyards in Dacia. Dacia was a kingdom north of the lower Danube in the first century BC and first century AD. -Dalby

The thorns which I have reaped

Are of the tree I planted.

They have torn me and I bleed:

I should have known what fruit

Would spring from such a seed.

-**Lord Byron**, *Childe Harold's Pilgrimage*

Erring Through Wine

But they also have erred through wine, and through intoxicating drink are out of the way; the priest and the prophet have erred through intoxicating drink, they are swallowed up by wine, they are out of the way through intoxicating drink; they err in vision, they stumble in judgment. -Isaiah 28:7

Franklin Graham on Alcohol

At a *Promise Keeper's Convention* Evangelist Franklin Graham spoke of his views on alcohol. For a long time he refused to give up drinking and used all the arguments for a Christian being able to drink. But he admitted there finally came a time when God convicted him to give up alcohol. He did so, and told the hundreds of men gathered for the meeting to stay away from it; have nothing to do with it.

"For the introduction of wine is perilous...An intoxicated woman is great wrath." -**Clement of Alexandria** (AD 150-215). an early Christian theologian.

Valentine on Noah

"Since Noah first grew grapes, made wine, passed out, and brought shame to himself and his family, the human race has been grappling with the moral dimension of the alcohol problem." -**Dr. Foy Valentine;** Baptist Press (bpnews.net), July 24, 2006. Valentine, a graduate of SWBTS, served as president of the *Christian Life Commission* from 1960-1987.

"Alcohol's drugging, depressing effect reduces mental capacity and thereby deadens moral sensitivity." -**Dr. Foy Valentine;** Baptist Press (bpnews.net), July 24, 2006.

"Choose well; your choice is brief and yet endless." -**Goethe**

Quitting, by George W. Bush

The first chapter in President George W. Bush's book, *Decision Points*, is entitled, "Quitting." His wife Laura asked him a simple question, "Can you remember the last day you didn't have a drink?" He replied, "Of course I can," but as he racked his brain, he could not remember one.

As a heavy drinker, the day after his 40th birthday, he gave up alcohol. "For months I had been praying that God would show me how to better reflect His will. My Scripture readings had clarified the nature of temptation and the reality that the love of earthly pleasures could replace the love of God…The booze was leading me to put myself ahead of others, especially my family…I knew I could count on the grace of God to help me change." Bush concluded, "Quitting drinking was one of the toughest decisions I have ever made. Without it, none of the others that follow in this book would have been possible." George W. Bush has acknowledged that had he not quit drinking, he would never have been President.

Altered by Alcohol, Controlled by the Holy Spirit?

"I don't believe there is a need for a foreign substance to achieve peace or relaxation, or whatever state some assert the moderate use of alcohol produces. Can a mind, altered by the consumption of alcohol, also be under the control of the Holy Spirit?" **-Dr. Richard Land,** president of the Ethics and Religious Liberty Commission of the Southern Baptist Convention. Baptist Press (bpnews.net), July 24, 2006.

"I have enough things in my life that distract me from my calling. I can't imagine intentionally ingesting a substance that will impair my judgment and further distract me from God's will for my life." **-Dr. Richard Land**, ERLC

Half the Legal Limit

"It appears it does not take much alcohol to impact a person physiologically. Researchers at the University of Washington in Seattle determined among test subjects that even one 'strong drink' can cause a 'substantial perceptual deficit.' This 'inattentional blindness' in those whose blood alcohol level was less than half the legal limit resulted in these individuals being more likely not to notice an object that appeared unexpectedly in their line of sight (Reuters, "One strong drink can make you 'blind drunk'," July 4, 2006)." -**Dr. Richard Land**, ERLC

Influence

"Most of the time we never know about the long-term payout of our actions. Who's watching? What are they seeing? We constantly are sending messages to those who are watching and listening. And as an ambassador of Christ (2 Corinthians 5:20), it behooves us all to consider carefully our ways." **-Dr. Richard Land,** ERLC

The Other End of the Beer Bottle

On a visit to New York City, John Bunyan Wilder and his wife and kids decided to attend Sunday morning services at the Jerry McCauley Water Street Mission in a bad part of town. As they walked through the streets, "I saw a pitiable object. A young man lay unconscious in a door-way. His clothing was utterly filthy and his hands were outstretched and grimy. His dark hair, once a rich black, was unkempt and his beard several days old. Flies crawled over him with no fear of being swatted. By his side lay a bottle that had contained some kind of cheap alcohol. He was dead drunk." "My wife hurried on but I spoke to my boys. 'Come here fellows,' I said. 'Here is the other end of the beer bottle. Here is what lies at the bottom of a glass of wine. This is the other end of the beer ad.' In words to this effect I addressed my sons. God help them never to forget the scene." -**John Bunyan Wilder**, *Stories To Live By*, Zondervan; 1964. (Deuteronomy 4:9)

God is Pleased When People Abstain

"Civil and spiritual leaders are told to abstain from alcohol as a beverage (Proverbs 31:4; 1 Timothy 3:3). Daniel refused to drink wine and God blessed his conviction (Daniel 1:8).

Evidence abounds that God is pleased when people avoid alcohol as a beverage. His blessing can and will fall upon those who refrain from imbibing." -**Dr. Jim Richards**, *Southern Baptist Texan*, July 31, 2006.

"A believer in no way can justify drinking if thereby he is contributing to the sustenance of an industry responsible for two-thirds of the violent deaths, two-fifths of all divorces, one-third of all crime, and untold millions of dollars in damage to private property. Such would violate all laws in the Bible, and especially the Corinthian principles." -**Paige Patterson**, president Southwestern Baptist Theological Seminary.

Why Lincoln Refused to Smoke or Drink

One day Abraham Lincoln was riding in a stage coach with a Kentucky colonel. The colonel took a bottle of whiskey out of his pocket and said, "Mr. Lincoln, won't you take a drink with me?"

"No, Colonel, thank you," replied Lincoln, "I never drink whiskey."

After a few more miles of pleasant conversation the Colonel reached in his pocket and brought out some cigars, saying, "Now, Mr. Lincoln, if you won't take a drink with me, won't you take a smoke with me? For here are some of Kentucky's finest cigars."

"Now, Colonel," said Lincoln, "you are such a fine, agreeable man to travel with, maybe I ought to take a smoke with you. But before I do, let me tell you a story of when I was a boy."

"My mother called me to her bed one day, when I was about nine years old. She was sick - very sick - and she said to me, 'Abey, the doctor tells me that I am not going to get well. I want you to promise me before I go that you will never use whiskey nor tobacco as long as you live.' And I promised my mother I never would. And up to this hour, Colonel, I have kept that promise. Now would you advise me to break that promise to my angel mother and take a smoke with you?"

The Colonel put his hand on Mr. Lincoln's shoulder and said with a trembling voice, "No, Mr. Lincoln, I wouldn't have you do it for the world. It was one of the best promises you ever made. I would give a thousand dollars today if I had made my mother a promise like that and had kept it as you have done." -*The Dry Legion*; **Walter B. Knight**.

"I am the sworn, eternal and uncompromising enemy of the liquor traffic. I have been, and will go on, fighting that damnable, dirty, rotten business with all the power at my command. I shall ask no quarter from that gang, and they shall get none from me. After all is said that can be said on the liquor traffic, its influence is degrading on the individual, the family, politics and business and upon everything that you touch in this old world. For the time has long gone by when there is any ground for arguments of its ill effects." -**Evangelist Billy Sunday** (1862-1935), professional baseball athlete, well known evangelist of the early 1900s.

"It is my opinion that the saloonkeeper is worse than a thief and a murderer. The ordinary thief steals only your money, but the saloonkeeper steals your honor and your character. The ordinary murderer takes your life, but the saloonkeeper murders your soul." **-Billy Sunday**

It's the Brain That Counts

"You can get along with a wooden leg, but you can't get along with a wooden head. It is the brain that counts, but in order that your brain may be kept clear you must keep your body fit and well. That cannot be done if one drinks liquor. A man who has to drag around a habit that is a danger and a menace to society ought to go off to the woods and live alone. We do not tolerate the obvious use of morphine or cocaine or opium and we should not tolerate intoxicating liquor because I tell you these things are what break down the command of the individual over his own life and his own destiny." **-Dr. Charles Mayo**, surgeon, addressing a large convention of boys. From Walter B. Knight.

"As a child, I realized the serious danger in alcohol use. Its addictive power is beyond debate. However, as a modern day, evangelical believer, I feel that it is very important to example and to teach abstinence regarding the use of alcohol." **-Dr. Frank S. Page**, SBC President, Executive Committee president.

Referring to alcohol and the 2006 SBC resolution, "We are not going to have people on our boards of trustees that do not believe in total abstinence…God saved me from that and when we make decisions, I want it to be done in sobriety. We may not always be right, but we'll always be sober." **-Executive Director John Sullivan**, Florida Baptist Convention, BP, 11-30-2006.

Surface Slush

"Persons seem content with perpetuating surface slush gleaned from a simplistic 'concordance' run on texts rather than serious interaction." **-Dr. Peter Lumpkins** on arguments of social drinkers who do not seriously engage the issue of the Bible and wine.

"When a person is inebriated, they lose inhibition, they lose self-control, and, for me, that's enough to say, 'Good thing to stay away from.'"
-**Dr. Jerry Sutton,** SBC Pastor's Conference.

Luther Burbank (AD 1849-1926) was a renowned horticulturalist, one of the most famous men of the early 1900s. He was also a liberal Unitarian. His mother Emma took a temperance pledge when she was 13 and kept it throughout her life. She apparently taught this conviction to her son.

Luther wrote, "I made a vow... that I would not touch a drop of any kind of liquor, and I shall keep it." He did keep that vow. He noted that almost all men who do not drink, are leading and respected men. He was a lifetime, honorary member of the *Woman's Christian Temperance Union,* although they were dismayed by his unorthodox Christianity. Upon being shown a new grape by Burbank, a visitor asked if he had improved the wine grape. Burbank said no, "Tell the children that I have never produced a superior wine grape and that if I ever do produce one, it shall be at once destroyed." -information from Dreyer; Knight

Southern Baptist Stand on Alcohol

"While the Bible may be subject to various interpretations concerning alcohol consumption (as well as the nature of the beverage consumed), Southern Baptists' understanding of the issue has been exceedingly unambiguous. In the Convention's history, SBC messengers have adopted over forty resolutions on the issue prior to this year—as recently as 2006 and as early as 1886." -**Dr. Richard Land and Barrett Duke,** *On Alcohol Use,* ERLC. erlc.com

Southern Baptists meeting in session have called their brothers and sisters to live "an exemplary Christian lifestyle of abstinence from beverage alcohol and all other harmful drugs" (1984); to recognize alcohol as "America's number one drug problem" (1982); to "reaffirm our historic position as opposing alcohol as a beverage" (1978); to view "personal abstinence" as the "Christian way" (1957); to express their "unceasing opposition to the manufacture, sale and use of alcoholic beverages" (1955); to realize alcohol is a "habit-forming and destructive poison" (1940) and the "chief source of vice, crime, poverty and degradation" (1936); and to "reassert our truceless and uncompromising hostility to the manufacture, sale, importation and transportation of alcoholic beverages" (1896).-**Dr. Richard Land and Barrett Duke**, *On Alcohol Use*, ERLC; 2006.

Pliny's Condemnation of Intoxicating Wine

"There is no department of man's life on which more labour is spent - as if nature had not given us the most healthy of beverages to drink, water, which all other animals make use of, whereas we compel even our beasts of burden to drink wine! And so much toil and labour and outlay is paid as the price of a thing that perverts men's minds and produces madness, having caused the commission of thousands of crimes, and being so attractive that a large part of mankind knows of nothing else worth living for!...

"The most cautious of these topers we see getting themselves boiled in hot baths and being carried out of the bathroom unconscious, and others actually unable to wait to get to the dinner table, no, not even to put their clothes on, but straight away on the spot, while still naked and panting, they snatch up huge vessels as if to show off their strength, and pour down the whole of the contents, so as to drink another draught; and they do this a second and a third time, as if they were born for the purpose of wasting wine, and as if it were impossible for the liquor to be poured away unless by using the human body as a funnel."

"Drunkards never see the rising sun, and so shorten their lives. Tippling brings a pale face and hanging cheeks, sore eyes, shaky hands that spill the contents of vessels when they are full, and the condign punishment of haunted sleep and restless nights, and the crowning reward of drunkenness, monstrous licentiousness and delight in iniquity. Next day the breath reeks of the wine-cask, and everything is forgotten - the memory is dead. This is what they call 'snatching life as it comes!' when, whereas other men daily lose their yesterdays, these people lose tomorrow also."[412] Pliny sometimes spoke approvingly of intoxicating wine, but his denunciation of it is remarkable.

Philo (20 BC - AD 50) was a Jewish writer living in Alexandria, Egypt. He says, "In abstemious men all the parts of the body are more elastic, more active and pliable, the external senses are clearer and less obscured, and the mind is gifted with acuter perception. The use of wine...leaves none of our faculties free and unembarrassed; but is a hindrance to every one of the them, so as to impede the attaining of that object for which each was fitted by nature...Speaking of the ascetic sect of Therapeutae, [Philo] says, 'They abstain from it (wine) because they regard it a sort of poison that leads men into madness.' On 'Drunkenness' he cites the case of Noah...and says 'It is evident that unmixed wine is poison.' Alluding to Aaron's name as indicating 'loftiness of thought,' he says, 'No one thus disposed will ever voluntarily touch unmixed wine or any other drug (pharmakon) of folly.'" [413]

Sam Jones and Tom Ryman

"One of the most dramatic of [Sam] Jones's many converts to Christianity was Tom Ryman of Nashville, a wealthy businessman and the owner of dozens of riverboats. At his tent meeting in Nashville, Jones railed against the drinking, gambling, and showgirls aboard these 'floating dens of iniquity.' Ryman had gone to the meeting intending to confront Jones, but by the end of the evening, Ryman confessed that Jones had 'whipped me with the gospel of Christ.' Ryman embraced Christianity and ceased selling liquor on his boats, the best of which he rechristened the *Sam Jones*. The converted businessman also paid to have a meetinghouse constructed for the use of churches and religious speakers

- an auditorium that would eventually become known as the Grand Ole Opry." -Joe L. Coker

Moral and Legal Suasion

"Some say use moral suasion against the saloon, and tell their boys to keep away from the saloon. That is good, but let us have legal suasion too, which will keep the saloon away from the boys. Let us have both kinds of suasion." -**Edgar Estes Folk,** quoted by Coker. Folk was a Tennessee Baptist leader, editor of *Baptist and Reflector*, president of the Southern Baptist Sunday School Board.

Toward the end of Prohibition, Franklin Roosevelt said saloons would not be allowed to reopen. Saloons had gotten a bad name for obvious reasons, and because of the Anti-Saloon League temperance organization. Instead, it seems that they reopened after Prohibition, but just did not call their businesses by the name saloon. That seems to be the reason there are very few "saloons" today, but plenty of clubs, beer joints, ice houses, and other liquor outlets.

Intoxicate

Ever notice the roots of the word intoxicate? It comes from the Latin meaning "in" and toxicum, "poison." The Greek root word is toxikon, a "poison in which arrows were dipped." -*Living Webster Encyclopedic Dictionary*; 1971. To become intoxicated from alcohol is to ingest the poison of alcohol. Never forget, alcohol is poison.

There Always Seemed to be at Least a Minority of Christians Who Abstained.

In writings once attributed to Justin, but now called Pseudo-Justin, "Wine is not to be drunk daily as water...Water is necessary; but wine only as a medicine...It is admitted that wine is a deadly poison...Church historian **Eusebius** speaks of abstinence from wine by (not all) Greeks, Romans, OT, NT, early church, including Timothy." [414] Dr. G. W. Samson continues by pointing out some, like the Ebionites, Tatians, Manichees, in the early church went so far as not to even have unfermented wine for the Lord's Supper. "We would certainly disagree with their practice, but still, it shows their strong conviction against intoxicating wine." [415]

Clement of Alexandria, c. AD 200 said, "I admire those who require no other beverage than water, avoiding wine as they do fire. From its use arise excessive desires and licentious conduct." [416] Clement was not an abstainer, but admired those who were. This reveals there were apparently many abstainers in the early church years.

Josiah Smith, a Presbyterian minister, published a sermon against alcohol in what is now the state of South Carolina in 1730. It was entitled, "*Solomon's Caution Against the Cup: A Sermon Delivered at Cainhoy, in the Province of South-Carolina.*" -from Coker

Catholics and Temperance

Catholics have generally not been friends of Temperance and use fermented wine for Communion or Mass. But there have been notable exceptions. The Olivetans, a Catholic group founded in AD 1319, "Were also fanatical total abstainers; not only was St. Benedict's kindly concession of a hemina of wine rejected, but the vineyards were rooted up and the wine-presses and vessels destroyed." -**Catholic Encyclopedia**; 1917. Now those were some strong abstainers!

Father Matthew, (1790-1856), was a famous Irish Catholic Priest who preached abstinence from alcohol. He very effectively promoted temperance and the temperance pledge in Ireland, England, and America. Total pledge takers from his ministry totaled in the hundreds of thousands. Statues of Father Matthew are in Cork and Dublin, Ireland.

Alcohol More Dangerous Than Heroin

LONDON -- Alcohol is more dangerous than illegal drugs like heroin and crack cocaine, according to a new study. British experts evaluated substances including alcohol, cocaine, heroin, ecstasy and marijuana, ranking them based on how destructive they are to the individual who takes them and to society as a whole. -**AP**, October 31, 2010; from published article in the medical journal, **Lancet**. For years preachers have been saying alcohol is the most dangerous drug. Here is confirmation of what they have been preaching.

Temperance

"Temperance may be defined as: moderation in all things healthful; total abstinence from all things harmful." -**Xenophon** (Greek philosopher), 400 BC. From *Women's Christian Temperance Union*.

An Empty Glass

An empty glass before the youth

Soon drew the waiter near.

"What will you take," the waiter asked,

"Wine red, or white, or beer?"

We've rich supplies of foreign brew

And wine your thirst to slake.

The youth with innocence replied,

"I'll take what my dad takes."

Swift as an arrow went the words

Into his father's ear,

And soon a conflict deep and strong

Awoke terrific fear.

Have I not seen the strongest fall,

The brightest led astray?

And shall I on my only son

Bestow a curse today?

Dad motioned to the waiter,

And gave his order clear;

I think I have a taste today,

For a sparkling glass of - iced tea!

-W. Hoyle; David R. Brumbelow; January 11, 2009

Great Poetry

Some temperance poems have not been the best of poetry. Hunt wrote a poem, "Cold Water Anthem." He admitted his temperance poem was not the best poetry. "I did not feel," said Hunt, "that I was a Burns or a Byron, but I did feel that I had made a poem that would have done Burns or Byron great good." -from Othniel A. Pendleton. I'll let you judge the previous poem.

"Alcohol is a product brewed with tears, thickened with blood, flavored with death."

"It is my judgment that because of the devastating problem that alcohol has become in America, it is better for Christians to be teetotalers except for medicinal purposes... The creeping paralysis of alcoholism is sapping our morals, wrecking our homes, and luring people away from the church." -**Evangelist Billy Graham**, *Christianity Today*, February 4, 1977.

D. L. Moody and Whiskey

"One Saturday evening he found in a house a jug of whiskey, which had been stored there for a carouse the following day. After a rousing temperance lecture, Mr. Moody persuaded the women of the house to permit him to pour the whiskey into the street. This he did before departing. Early the next morning he came back to fetch the children of the place to Sunday school. The men were lying in wait for him to thrash him. It was impossible to get away, for he was surrounded on all sides, but before they could touch him, Mr. Moody said, "See here, men, if you are going to whip me, you might at least give me time to say my prayers." The request was unusual; perhaps it was for that very reason that it was acceded to. Mr. Moody dropped upon his knees and prayed such a prayer as those rough men had never heard before. Gradually they became interested and then softened, and when he had finished they gave him their hands, and a few minutes later Mr. Moody left the house for his school, followed by the children he had come to find." [417] Moody was a great evangelist of the 1800s.

Temperance Movement - From Moderate Drinking to Abstinence

The modern day (if you can count the 1700s to the present, modern day) Temperance Movement began in the 1700s. At this time in America drinking was very common, even among Christian leaders. Drinking was accepted and abused. Ironically, the Temperance Movement began as a movement in favor of moderate drinking; only opposing drunkenness. It soon became apparent, however, that moderate drinking did not solve the problem; it exacerbated the problem. In countless cases, moderate drinking leads the drinker and others into immoderate drinking. So temperance leaders began to preach and teach abstinence from alcohol. While many Christians have abstained from alcohol throughout history, the church seemed to especially awake to the evils of alcohol during this time. Somewhat like slavery, what had once been condoned by many believers was now recognized to be condemned in Scripture.

Some social drinkers point out that Baptists and others drank on a regular basis two centuries ago. This is generally correct. Otherwise outstanding Christians were dead wrong on this issue. It seemed they had a huge blind spot, until awakened by the Temperance Movement. They tried moderate drinking, but that still caused diminished good judgment, led to drunkenness, and other problems. So they then began to preach and teach abstinence from beverage alcohol. The following quote reveals some of this attitude in the early 1800s.

Alcohol Common in Early 1800s

"Our honest forefathers...made an earnest effort to protect church members from the contaminations of the world. To this end they placed some worldly amusements under the ban of stern disapprobation, and made them subjects of discipline. In making these discriminations they made, we must admit, some grave mistakes. But I think it will be found that their mistakes were in allowing some things which they ought to have forbidden rather than in those things which they condemned. For instance, they condemned social dancing and card playing, because they judged these amusements to be hurtful to spiritual growth and danger-ously alluring to other and grosser vices. In this judgment they were cer-tainly right. But they allowed social drinking without a word of censure. In this they were inconsistent. But it should be observed that they were not knowingly inconsistent. I can distinctly remember when a moderate 'dram' (so called) was deemed as harmless as the same amount of milk. Not only so, it was considered, in hundreds of cases, to be actually help-ful and sustaining to the physical system. Hence the most prudent farm-ers of the country would often furnish it to their hands, especially when their work happened to be more than usually heavy. Religious people did the same thing. Even preachers, after a long sermon, would often relieve their fatigue with a dram.

In the light of these facts it is not surprising that our fathers, seventy years ago, should overlook the evils of moderate drinking. But remember that they did not tolerate drunkenness. Their church minutes show many cases of discipline, which evince their watchfulness over the morals of their brethren." -**S. G. Hillyer**, [418]

John Dagg on Baptists & Alcohol in Early 1800s

"In August, 1812, I attended the meeting of the Ketocton Association, to which our church belonged; and was distressed to see the free use made of ardent spirits, by the ministers and members. There was also distressing evidence, that the principal deacon of our church indulged freely in the use of the pernicious liquor; though we had no proof that he was guilty of gross drunkenness. These facts induced me to prepare a query, which the church, at my request, sent up to the Association, at its next meeting. 'At what point between total abstinence from ardent spirits, and intoxication by them, does the use of them become sinful?' The temperance reform was then unknown, and the notion of total abstinence was so little understood, that the bearing of my query was not apprehended. In replying to it, the Association replied, that *moderation* was necessary in the use of ardent spirits. This was the doctrine of the times, in which multitudes of Christian professors, including ministers of the gospel, were victims of intemperance. The deacon just referred to, I assisted afterwards, to exclude from the church; and, some time after, while lying on his hearth, in a state of intoxication, he was roasted to death by the fire."

"Soon after my settlement in Philadelphia, it became necessary to give a practical proof of my opposition to the use of ardent spirits. The ministers of the Association were accustomed to meet every three months at some one of the churches. A sermon was delivered by a brother appointed at the previous meeting. After the sermon, the ministers dined with the pastor; and, in the afternoon, in a ministerial conference, criticized the sermon for the common benefit. In the first meeting of this sort that I attended, my heart was pained to see ardent spirits set out on the pastor's side-board, and the guests partaking freely. At subsequent meetings the same custom was observed. At length it became my turn to entertain the ministers meeting. The best food that the market afforded, I gladly provided for the table; but my conscience would not permit me, to offer the pernicious beverage. The effect, I think, was good. So far as I know, the decanter was never seen afterwards at a minister's meeting." [419] With the widespread abuse of alcohol during the 1700s and 1800s, thank God for the Temperance movement.

The Courage To Say No

You're starting, my son, on life's journey,

Along the grand highway of life.

You'll meet with a thousand temptations-

Each city with evil is rife.

This world is a stage of excitement-

There's danger wherever you go;

But if you are tempted in weakness,

Have courage, my son, to say No!

Be careful in choosing companions,

Seek only the brave and the true;

And stand by your friends when in trial,

Ne'er changing the old for the new.

And when by false friends you are tempted,

The taste of the wine cup to know,

With firmness, with patience, and kindness,

Have courage, my son, to say No!

-unknown

Charles H. Spurgeon on Alcohol

Occasionally you hear someone defending social drinking by using Charles Haddon Spurgeon (AD 1834-1892) as an example, or as an excuse. If the great Baptist preacher of London did not believe in abstinence from beverage alcohol, then it must be alright for us to drink. Contrary to this view, we should follow biblical teaching and the common sense God gave us. Some will be surprised, however, to see what Spurgeon came to believe about the recreational use of the drug alcohol. In his early years, Charles Haddon Spurgeon disagreed with those who preached abstinence from alcohol. But as time went by, the temperance advocates (those who promoted total abstinence from beverage alcohol), convinced Spurgeon; or maybe the Holy Spirit convinced him. Spurgeon actually became a temperance advocate. Temperance meetings were held in Spurgeon's church, the Metropolitan Tabernacle. In 1882 Spurgeon would boldly declare, "Next to the preaching of the Gospel, the most necessary thing to be done in England is to induce our people to become abstainers." I'm strongly against drinking, but I'm not even sure that I would go as far as Spurgeon, with his above statement. Dr. Lewis Drummond stated of Spurgeon, "Obviously he had become a strong advocate for abstinence." The above two quotes are from *Spurgeon: Prince of Preachers* by Lewis A. Drummond, Kregel Publications; 1992. Dr. Drummond was a professor at Southern Baptist Theological Seminary and president of Southeastern Baptist Theological Seminary. Charles Spurgeon also spoke to the issue of Communion

Wine stating that his church used only unfermented wine. "We use Frank Wright's unfermented wine at the Tabernacle, and have never used any other unfermented wine. I am given to understand that some of the so-called unfermented wine has in it a considerable amount of alcohol; but Mr. Wright's is the pure juice of the grape. One person advertised his wine as used at the Tabernacle though we had never used it even on one occasion. So far as we are concerned, we use no wine but that produced by Messrs. Frank Wright, Mundy, and Co. Yours truly, **C. H. SPURGEON.**" [420] Notice that Spurgeon calls this pure juice of the grape "unfermented wine."

"I hope they will be full of spirit against evil spirits, stout against stout, and hale against ale." -**Charles H. Spurgeon**, letter to temperance society, March 19, 1884. Next time you hear someone say Spurgeon was for social drinking, tell them the rest of the story. Charles H. Spurgeon became a strong advocate of total abstinence from alcohol.

Jack and Jill

Went up the hill,

To get some bootleg liquor;

Jack went blind

And lost his mind -

And Jill is even sicker!

-**unknown**, *quoted by R. G. Lee*

The Pledge

signed by Frances R. Willard when a child

"A pledge we make, no wine to take,

Nor brandy red that turns the head,

Nor fiery rum that ruins home,

Nor whiskey hot that makes the sot,

Nor brewer's beer, for that we fear;

And cider, too, will never do-

To quench our thirst we'll always bring

Cold water from the well or spring;

So here we pledge perpetual hate

To all that can intoxicate."

Signal Press - Evanston, Ill.
-c. 1920s.

"Temperance will do you no harm. In a thousand ways, it will do you good. Even occasional drinking will do you no good; and entire abstinence from drinking will do you no harm." -**Rev. Henry Ward Beecher**, Plymouth Church, Brooklyn, NY. *Temperance Sermons*, National Temperance Society, 1873.

Lincoln on Slavery and Prohibition

"Less than a quarter of a century ago I predicted that the time would come when there would be neither a slave nor a drunkard in the land. I have lived to see, thank God, one of those prophecies fulfilled. I hope to see the other realized." -**President Abraham Lincoln**, on the last day of his life, speaking to Chaplin James B. Merwin. Referred to by Charles Thomas White and Walter B. Knight.

"Drunkenness is simply voluntary insanity." -**Seneca**, *Epistolae Morales*; c. AD 60.

"One of the disadvantages of wine is that it makes a man mistake words for thoughts." **Samuel Johnson**

A Drunk Doctor

I never knew my grandmother. She died when my dad, Joe Brumbelow, was 6 years old. In 1936 Carrie Mae Brumbelow became seriously ill with appendicitis. Her husband, E. P. Brumbelow took her to the nearest hospital in Sugar Land, Texas. Once there, the doctor could not be found. They tried to contact him to no avail. In desperation, E. P. went looking for him. He finally found the medical doctor drunk in a beer joint. By the time E. P. got back to the hospital, Carrie Mae was dead. Grandpa used to say, "I never wanted to kill a man so bad in all my life, as I wanted to kill that doctor that day." TV sometimes shows how funny and amusing it is to see medical doctors make their own brew and get drunk. But what happens when a medical doctor is drunk and there is a medical emergency? There is nothing funny about alcohol when it has cost the life of your innocent wife.

Southern Baptist Resolution against Alcohol

On Alcohol Use in America; June 2006

WHEREAS, Years of research confirm biblical warnings that alcohol use leads to physical, mental, and emotional damage (e.g., Proverbs 23:29-35); and WHEREAS, Alcohol use has led to countless injuries and deaths on our nation's highways; and WHEREAS, The breakup of families and homes can be directly and indirectly attributed to alcohol use by one or more members of a family; and WHEREAS, The use of alcohol as a recreational beverage has been shown to lead individuals down a path of addiction to alcohol and toward the use of other kinds of drugs, both legal and illegal; and WHEREAS, There are some religious leaders who are now advocating the consumption of alcoholic beverages based on a misinterpretation of the doctrine of "our freedom in Christ"; now, therefore, be it RESOLVED, That the messengers to the Southern Baptist Convention meeting in Greensboro, North Carolina, June 13-14, 2006, express our total opposition to the manufacturing, advertising, distributing, and consuming of alcoholic beverages; and be it further

RESOLVED, That we urge that no one be elected to serve as a trustee or member of any entity or committee of the Southern Baptist Convention that is a user of alcoholic beverages. RESOLVED, That we urge Southern Baptists to take an active role in supporting legislation that is intended to curb alcohol use in our communities and nation; and be it further RESOLVED, That we urge Southern Baptists to be actively involved in educating students and adults concerning the destructive nature of alcoholic beverages; and be it finally RESOLVED, That we commend organizations and ministries that treat alcohol-related problems from a biblical perspective and promote abstinence and encourage local churches to begin and/or support such biblically-based ministries. -sbc.net

Assembly of God Statement on Alcohol

"The Christian who advocates or condones 'drinking in moderation' is providing Satan an opening he would not have with an individual committed to total abstinence. By medical definition, alcohol is a drug. The moderate drinker is naïve if he does not recognize the peril of eventually becoming addicted himself." -Abstinence, **Assemblies of God** Position Paper; August 6, 1985.

Beware Self Righteousness

"Beware of self righteousness. That man's problem is not his liquor; he needs Jesus. Don't go looking down your long nose at somebody because you don't drink. Some people are going to Hell perhaps because they drink. Some people are going to Hell because they never drink; and they're so rotten self righteous they don't see their need of Jesus. If you never touched a drop of liquor in all of your life, and don't get saved, it'll just mean you'll go to Hell sober. You need Jesus. Trust in Christ and though your sins be as scarlet, they'll be white as snow. Though they be red like crimson, they shall be as wool." -**Adrian Rogers**, *The Battle of the Bottle*.

Maston on Common Sense

"Some people have more of a weakness for alcohol than others, but 'No one can tell who can indulge without overindulging.' Dr. E. M. Jellinek, generally considered America's leading authority on alcohol, suggests: 'Science has found no way to determine or distinguish who or what sort of a drinker may or may not become an alcoholic.' Is it not just plain common sense for one to abstain entirely from all forms of beverage alcohol? 'At the last it bites like a serpent, and stings like an adder.'" -**Dr. T. B. Maston**, *Right Or Wrong?*, Broadman Press, 1955.

Maston was a longtime Ethics Professor at Southwestern Baptist Theological Seminary, Fort Worth, Texas.

"A majority of the 100 million drinkers in America today are church-goers who have been taught that the Bible sanctions a moderate use of alcoholic beverages. Unfortunately, moderate drinking has led over 18 million Americans to become immoderate drinkers, because alcohol is a habit-forming narcotic that weakens one's capacity for self-control." -**J. Gerald Harris**, Editor, *Christian Index*, Georgia; 10-21-2004.

"Alcohol in beer, wine or whisky - legal or bootleg - is a habit-forming narcotic, poison that befuddles the brain, depresses the nerves, and releases inhibitions. Men who, when sober, are kind, tender fathers and husbands will, when under the influence of alcohol, become overbearing, brutal, tyrants of their homes. Men, who when sober, are upright and law-abiding citizens, get under the influence of booze and become lawbreakers and crooks committing the most heinous crimes." -**Sam Morris**, *The Voice of Temperance*, San Antonio, TX; c. 1947.

Rum's Doings in Early Texas

"Rev. W. Y. Allen; in *Allen's Reminiscences of Texas, 1838-1842*, mentioned an address by President Sam Houston to Congress, adding that it was followed by 'a fight in front of the Capital and a murder in the after-noon...The murderer and murdered were both heroes of San Jacinto - rum's doings.'" At that time the capital was in Houston, and Texas was an independent nation. San Jacinto referred to the famous 1836 Battle at San Jacinto that won Texas' independence. Another of many sad accounts of lives ruined by alcohol.

Medical Marijuana

"People do not want to see someone suffer from illness, so they are prepared to approve measures they are told will bring relief. Unfortunately, medical marijuana does not deserve that kind of response. Marijuana is a hallucinogenic and extremely debilitating. People who use marijuana to help them deal with severe pain must also use other pain-relieving drugs because marijuana is not that effective in relieving severe pain. The primary contribution medical marijuana will make to society is that it will decrease society's legitimate concern about the dangers of marijuana use and lead to greater acceptance of the drug. Medical marijuana causes people to drop their guard and sets them up for broader legalization efforts. That's why legalization of medical marijuana almost always precedes efforts to legalize the recreational use of marijuana."-**Dr. Barrett Duke**, Ethics & Religious Liberty Commission, BP 8-17-2010.

A Man and a Dog

"I have seen a man and a dog go into a saloon, and in an hour the man would get beastly drunk and stagger out like a hog, while the dog would come out and walk away like a gentle-man." **Sam Jones** (1847-1906), outstanding Methodist evangelist. "Wisdom is beclouded by wine." **-old Roman proverb**

"No amount of alcohol is safe to drink during pregnancy." -**Centers for Disease Control**; 2010.

A Word to the Reader

It is high time that Christians start speaking out knowledgeably about wine, alcohol, and the Bible. Those writing study Bibles, commentaries, and Sunday School lessons need to be aware of this information. Please feel free to use the quotes and material herein to speak out and put in print this message. Speak out on the blogs and the internet. Leave comments with this information. Let your pastor and youth minister know the truth about the Bible and wine.

A simple note such as the following could be placed in study Bibles and Bible dictionaries. It could clear up a lot of confusion.

Wine - in Bible and ancient times referred to both alcoholic and non alcoholic wine. Aristotle, Hippocrates and others said sweet wine would not inebriate (fermentation took away the sweetness). Ancients also had the knowledge to easily preserve wine in a nonalcoholic state. Interpret the kind of wine by the context.

Shekar - sometimes translated similar drink, strong drink, or beer. Many authorities say shekar could be either an intoxicating drink, or non-intoxicating sweet drink, from fruit other than the grape. Or a shorter note:

Wine - referred to both intoxicating (Proverbs 20:1; 23:29-35) and un-intoxicating (Proverbs 3:10; Isaiah 16:10; 65:8) drinks.

The Bridge Builder

An old man, going a lone highway,

Came at the evening, cold and gray,

To a chasm, vast and deep and wide,

Through which was flowing a sullen tide.

The old man crossed in the twilight dim-

That sullen stream had no fears for him;

But he turned, when he reached the other side,

And built a bridge to span the tide.

"Old man," said a fellow pilgrim near,

"You are wasting strength in building here,

Your journey will end with the ending day;

You never again must pass this way.

You have crossed the chasm, deep and wide,

Why build you the bridge at the eventide?"

The builder lifted his old gray head.

"Good friend, in the path I have come," he said,

"There followeth after me today

A youth whose feet must pass this way.

This chasm that has been nought to me

To that fair-haired youth may a pitfall be.

He, too, must cross in the twilight dim;

Good friend, I am building the bridge for him."

-Will Allen Dromgoole

APPENDIX I

Books on the Bible and Wine
Taking an Abstinent Position

Some books are modern and in print. Some are older but have been reprinted and are available. Some are on the internet at googlebooks, etc. Check with your local bookseller to order, or do a search on amazon.com or alibris.com. While disagreeing with some interpretations and details, the following are worthwhile books from the abstinence point of view.

Anderson, Jim, *A Biblical Study on Wine*, Anderson Evangelistic Enterprises, 303 Hargis Court, Belton, MO 64012; 1980. Pamphlet by professor at MWBTS.

Bacchiocchi, Samuele, *Wine in the Bible*, Biblical Perspectives, Michigan; 2004. Good abstinence book from an Adventist scholar. Refutes some abstinence critics.

Baptist Press, Nashville, TN; bpnews.net. BP, news agency for the Southern Baptist Convention, has published a number of excellent articles on alcohol. Do a search on their website.

Barnes, Albert, *Barnes' Notes*, Baker Book House, Grand Rapids, Michigan; 1885.

Bengay, George Washington, *Temperance Anecdotes, Original and Selected*, National Temperance Society, 58 Reade Street, New York; 1873.

Criswell, W. A., wacriswell.com. Sermons by pastor who stood strong against alcohol.

Delavan, Edward C., *Temperance Essays*, National Temperance Society, 172 William Street, New York; 1869.

Ethics and Religious Liberty Commission, SBC, Nashville & Washington, DC; erlc.com. Good information on issues such as drinking, gambling, pro-life, religious liberty.

Ewing, Charles Wesley, *The Bible and Its Wines*; 1985. Very good, presenting a wealth of evidence. Should be available from amazon.com.

Field, Leon C., *Oinos: A Discussion of the Bible Wine Question*, Phillips & Hunt, New York; 1883. Reprinted by Kessinger Publishing. One of the best books on the Bible and wine.

Geisler, Norman L., *To Drink or Not to Drink*; www.normgeisler.com.

Gordon, Anna A., *The Beautiful Life of Francis E. Willard*, Woman's Temperance Publishing Association, Chicago, Illinois; 1898. Famous temperance advocate.

Gordon, Ernest Barron, *Christ, The Apostles and Wine*, Sunday School Times; 1944. Good brief study. Ernest Gordon was the son of A. J. Gordon. Book has been reprinted.

Hearn, C. Aubrey, *Alcohol and Christian Influence*, Convention Press, Nashville, Tennessee, 1957. Very good book, worth reprinting.

Impe, Jack Van and Campbell, Roger F., *Alcohol, The Beloved Enemy*, Jack Van Impe Ministries, Troy, Michigan; 1980. Good, solid book on drinking.

Kerr, Norman Shanks, *Wines, Scriptural and Ecclesiastical*, National Temperance Publication Depot, London; 1882. Excellent, brief book. Dr. Kerr (1834-1899) was a Christian medical doctor who extensively studied the issue of ancient wine. According to the *New York Times* (June 1, 1899), Dr. Kerr was considered the most skillful physician in England in treating alcoholism.

Kitto, John, *Popular Cyclopedia of Biblical Literature*, Fredonia Books, Amsterdam, The Netherlands; 1852. Has been reprinted. Good work highly praised by Spurgeon.

Knight, Walter B., Three Thousand Illustrations, Wm. B. Eerdmans Publishing, Grand Rapids, Michigan; 1947, 1971. Few illustration books today have abstinence information; this book does not make that error. It is under the heading *Temperance*.

Lee, R. G., *Highways to Havoc*, Christ for the World Publishers, Orlando, Florida; 1974. Great preacher who often spoke against alcohol.

Lees, Frederic Richard and Burns, Dawson, *The Temperance Bible Commentary*, National Temperance Society and Publication House, New York; 1870. A monumental work dealing with most every biblical reference to drinking. A wealth of scholarly evidence. I have an original copy. Available online. A few original copies still available. Should be reprinted; if you would like to help do so, contact Brumbelow or Free Church Press.

London, Jack, *John Barleycorn*, Quill Pen, New York; 2008. Fascinating autobiography of the famous author Jack London and his ultimately loosing struggle with alcohol.

Lumpkins, Peter, *Alcohol Today*, Hannibal Books, Garland, TX; 2009. Excellent recent book presenting biblical as well as philosophical evidence against drinking. Foreword by Jerry Vines.

Masters, Peter, *Should Christians Drink?*, The Wakeman Trust, London; 2001. Good book from the pastor of the Metropolitan Tabernacle, London, England.

McGuiggan, Jim, *The Bible, the Saint, and the Liquor Industry*; 1977.

Morris, Sam, *The Woe of the Wine Cup*, The Voice of Temperance, San Antonio, Texas; 1937.

Morris, Sam, *Wine, Women and Song*, The Voice of Temperance, San Antonio, Texas; 1938.

Morris, Sam, *Scrap Book II*, The Voice of Temperance, San Antonio, Texas; 1940.

Morris, Sam, *National Prohibition*, The Voice of Temperance, San Antonio, Texas; c. 1947. Sam Morris was a faithful soldier in the fight against alcohol.

Nott, Eliphalet, *Lectures on Temperance*; 1847.

Patton, William, *Bible Wines*, 1871. One of the old classics. Re-printed and available today.

Pickett, Deets, Editor, *The Cyclopedia of Temperance, Prohibition and Public Morals*, Methodist Book Concern, New York; 1917.

Resolutions of the Southern Baptist Convention on Alcohol; sbc.net

Reynolds, Stephen M., *The Biblical Approach to Alcohol*, L. L. Reynolds Foundation, Glenside, PA; 2003. Reynolds had a Ph.D. from Princeton University and worked in biblical and oriental languages. Contributed to the translation of the NIV. Very good, scholarly book. Refutes some abstinence critics.

Rice, John R., *The Double Curse of Booze*, Sword of the Lord, Murfreesboro, Tennessee; 1960. Pamphlet, evangelistic sermon against alcohol.

Rogers, Adrian P., *The Battle for the Bottle*, parts 1&2; lwf.org, 877/568-3463. Excellent sermons on alcohol.

Samson, George Whitefield, *The Divine Law as to Wines*, National Temperance Society and Publication House, New York; 1880. Good, scholarly book by a University president.

Select Temperance Tracts, American Tract Society, 150 Nassau-Street, New York; c. 1870. Reprinted by Michigan Historical Reprint Series, University of Michigan. Includes a number of tracts from the early days of the temperance movement.

Sinclair, Upton, *The Cup of Fury*, Fleming H. Revell, Westwood, New Jersey; 1956. Riveting book against alcohol from a respected, prolific author and a liberal socialist. Tells of talented friends destroyed by alcohol.

Sumner, Robert L., *Fights I Didn't Start, And Some I Did*, Biblical Evangelism, 5717 Pine Drive, Raleigh, NC 27606; 2009. Good chapter on "Is God a Social Drinker?"

Sumner, Robert L., *The Biblical Evangelist*, 5717 Pine Drive, Raleigh, North Carolina 27606-8947; biblicalevangelist.org.

Parts of this book, Ancient Wine, were originally published in the *Biblical Evangelist*.

Sumner, Robert L., *The Blight of Booze*, Biblical Evangelism, 5717 Pine Drive, Raleigh, NC 27606; 1955. Excellent booklet / sermon preaching against alcohol.

Southern Baptist Texan, P.O. Box 1988, Grapevine, Texas 76099. texanonline.net. Excellent article, *Can you be a Biblical Inerrantist and Oppose the Use of Alcohol as a Beverage?* by Jim Richards; July 31, 2006.

Teachout, Richard, *On the Fruit of the Vine*, Bible Studies for Today, 8890, boul. Ste-Anne, Chateau-Richer, QC GOA1NO, Canada; 2010.

Teachout, Robert P., *The Use of Wine in the Old Testament*. Doctoral Dissertation at Dallas Theological Seminary, Dallas, TX; 1979. Available from UMI Dissertation Service, P.O. Box 1346, Ann Arbor, Michigan 48106; il.proquest.com. Rather technical but excellent, scholarly information on wine.

Teachout, Robert P., *Wine, The Biblical Imperative: Total Abstinence*; 1983, 1994. Order from: Dr. Robert Teachout, 22544 West Road, Apt. #204, Woodhaven, MI 48183. $10 postpaid.

Very good brief book. Condenses his doctoral dissertation and puts it on a more popular level.

Thayer, William M., *Communion Wine*, National Temperance Society and Publication House, New York; 1869. Another old classic. Reprinted by University of Michigan.

Vines, Jerry, *Libertinism: A Baptist and His Booze* by Jerry Vines; jerryvines.com. Good, solid sermon on this issue.

White, Charles T., *Lincoln and Prohibition*, Abingdon Press, New York; 1920. Introduction by Will H. Hays, Postmaster General of the United States. Charles T. White was political news editor of the New York Tribune and New York Commissioner.

Whitmore, Orin B., *Bible Wines vs. The Saloon Keeper's Bible*, Press of the Alaska Printing Co., Seattle, 1911.

Wilkerson, David, *Sipping Saints*, David Wilkerson Publications, Lindale, TX; 1978.

Women's Christian Temperance Union - wctu-sd.org; wctu.org. One of the oldest temperance organizations, still promoting abstinence.

Additional Books Researched on the Bible and Wine

Alcock, Joan P., *Food in the Ancient World*, Greenwood Press, Westport, Connecticut, London; 2006.

Anonymous, *The New Family Receipt Book: Containing 800 Truly Valuable Receipts*, John Murray, London; 1820. Recipes (receipts) and remedies of the early 1800s.

Barry, Sir Edward, *Observations Historical, Critical, and Medical, on the Wines of the Ancients. And the Analogy Between Them and Modern Wines. With General Observations on the Principles and Qualities of Water and in Particular on Those of Bath*, Fellow of the Royal College of Physicians, and of the Royal Society, T. Cadell; 1775.

Bennett, William J., *The Devaluing of America: The Fight for our Culture and our Children*, Simon & Schuster, New York; 1992.

Bottero, Jean, *The Oldest Cuisine in the World*: Cooking in Mesopotamia, University of Chicago, Chicago & London; 2004. Originally published in French.

Bubel, Mike & Nancy, *Root Cellaring*, Storey Publishing, MA; 1991.

Bush, George W., *Decision Points*, Crown Publishers, New York; 2010.

Classical Writers

Classical Period covered roughly 800 BC - AD 600.

Aristotle; Cato; Varro; Pliny; Columella; Athenaeus; Theophrastus; Hippocrates; Polybius; Homer; Plutarch; Chrysostom; Augustine; Josephus; and others. These writings are available from the *Loeb Classical Library*, other publishers, and several websites. Of course the Bible itself was written before and during the Classical Period. *Ancient Wine and the Bible* has used the *New King James Version* (NKJV) of Holy Scripture.

Internet Classics Archive - http://classics.mit.edu/index.html

Perseus Digital Library - http://www.perseus.tufts.edu/hopper/

Coker, Joe L., *Liquor in the Land of the Lost Cause*, The University Press of Kentucky, Lexington; 2007. Interesting historical account of temperance movement in Alabama, Georgia, Tennessee.

Cox, Jeff, *From Vines to Wines*, Storey Publishing, North Adams, MA; 1999.

Dalby, Andrew, *Food in the Ancient World From A-Z*, Routledge, London & New York; 2003. Dalby is a historian and linguist. Best book I've found on this subject. This is not a religious book, but it clearly reveals the fact of un-intoxicating wine in the ancient world. Arranged in encyclopedic form.

Dreyer, Peter, *A Gardener Touched With Genius: The Life Of Luther Burbank*, Luther Burbank Home & Gardens, Santa Rosa, California; 1985.

Evans, Hilary & Mary, *The Man Who Drew the Drunkard's Daughter: The Life and Art of George Cruikshank*, Frederick Muller Limited, London; 1978.

Fallon, Sally, *Nourishing Traditions*, New Trends Publishing, Washington, DC; 2001. Information on lactic fermentation.

Fur-Fish-Game magazine, furfishgame.com. This and other outdoor, gardening, farm, hunting, and fishing magazines occasionally have information on properly preparing and preserving food and drink.

Gardeners and Farmers of Terre Vivante, *Preserving Food Without Freezing or Canning: Traditional Techniques Using Salt, Oil, Sugar, Alcohol, Vinegar, Drying, Cold Storage, and Lactic Fermentation*, Chelsea Green Publishing Company, White River Junction, Vermont; 2006. Information on old time methods of preservation.

Greenberg, Florence, *Jewish Cookery*, Penguin Books; 1947, 1963.

Hunter, Beatrice Trum, *Fermented Foods and Beverages: An Old Tradition*, Keats Publishing, New Canaan, Connecticut; 1973.

Ivy, James D., *No Saloon in the Valley*, Baylor University Press, Waco, TX; 2003. History of temperance in Texas.

Kobler, John, *Ardent Spirits: the Rise and Fall of Prohibition*, Da Capo Press, New York; 1973. Good summary of temperance and prohibition movement by one not necessarily in favor.

Lehman's Non-Electric Catalog; lehmans.com. Sells to Amish and anyone else. Among other things, provides information on food preservation.

Lesko, Leonard H., *King Tut's Wine Cellar*, BC Scribe Publications, Berkley, CA; 1977.

Loubat, Alphonse, *The American Vine Dresser's Guide*, G. & C. Carwill, New York; 1827.

McGovern, Patrick E., *Ancient Wine: The Search for the Origins of Viniculture*, Princeton University Press, Princeton & Oxford; 2003.

Nathan, Joan, *Jewish Cooking in America*, Alfred A. Knopf, New York; 1994, 2001.

Pellechia, Thomas, *Wine: The 8,000 Year Old Story of the Wine Trade*, Running Press, Philadelphia, London; 2006.

Proulx, Annie, *Making the Best Apple Cider*, Storey Publishing, MA; 1980.

Robinson, Jancis and Harding, Julia, *Oxford Companion to Wine*, Oxford University Press, Oxford & New York; 2006.

Shephard, Sue, *Pickled, Potted, Canned: How the Art and Science of Food Preserving Changed the World*, Simon & Schuster, New York, London; 2000.

Stein, Robert H., *Difficult Passages in the New Testament*, Baker, Grand Rapids, Michigan; 1990.

Thayer, William M., *A Biography of Abraham Lincoln*; 1882.

The Pledge

"**The Pledge**" was often taken in days past concerning alcohol. Of course some did not take it seriously and later gave in to drinking. But some were encouraged by that pledge and took it very seriously. It also gave them a reply to those trying to entice them to drink, "I have taken a solemn vow to God, my family, and my church that I will not drink. Now, would you want me to break that vow?" Any true friend would never persuade you to do so.

An example of "The Pledge" is given below.

The Pledge

I solemnly promise, by the help of God,

* To abstain from the use of all intoxicating drinks as a beverage, and the recreational or unnecessary use of any drug.

* To honor God in my life, mind, and body.

Signed: _____

City/State: _____

Date: _____

Sweeter Than Wine

Suppose you lived two or three thousand years ago and had no access to sweets. People crave sweet things. Imagine the joy and satisfaction, after a year of diligent labor, of being able to drink fresh pressed, incredibly sweet, new wine. Wine so sweet it was syrupy. But there is something sweeter than wine. Sweeter by far is to experience forgiveness, be made right with God, and have everlasting life and a home in Heaven. It is sweet to find purpose and meaning to life. Please permit me to tell you how you may personally know God.

All mankind are sinners. Our sin separates us from the one true holy God.

All have sinned and fall short of the glory of God. -Romans 3:23

The final consequence of our sins is eternal death, separation, Hell. But in spite of our sin, God has an incredible gift to give us.

The wages of sin is death, but the gift of God is eternal life in Christ Jesus our Lord. -Romans 6:23

God's gift of eternal life costs you nothing. It cost God, however, the life of His only begotten Son.

God demonstrates His own love toward us, in that while we were still sinners, Christ died for us. -Romans 5:8

Jesus Christ, God the Son, loves you so much, He bled and died for you and literally rose again and is living today.

You cannot pay for God's gift. You can't work for it. But you do have to admit your sins to God and take or receive his gift by asking Him to be your Lord (Master, Boss) and Savior.

As many as received Him, to them He gave the right to become children of God, to those who believe in His name. -John 1:12

If you confess with your mouth the Lord Jesus and believe in your heart that God has raised Him from the dead, you will be saved. For with the heart one believes unto righteousness, and with the mouth confession is made unto salvation…Whoever calls on the name of the LORD shall be saved. -Romans 10:9-10,13.

Prayer:

Lord Jesus, I believe You died on the cross for my sins and rose again. I know I'm a sinner and have done wrong in Your eyes. Please forgive me of all my sins. Come into my heart and be my Lord and Savior. Help me to live for You each day. Thank You for loving and saving me. In Jesus' name, Amen.

Today I accepted Jesus Christ as my Lord and Savior:

Name:

City & State:

Date:

Once you've accepted Jesus as your Lord and Savior:

1. Find a good (none are perfect) Bible believing church and attend every Sunday morning, Sunday night, Wednesday night.

2. Follow the Lord in baptism.

3. Read a chapter in your Bible every day. Start in the book of John or Luke. Do what it says.

4. Pray to God every day. When you sin, ask His forgiveness.

5. Tell others about Jesus and His salvation.

6. Give to the Lord, His church, to missions.

7. Make Jesus the Lord of your life.

Enjoy the sweet wine of following Jesus and fellowship with His people; stay away from the bitter wine and drugs of the devil.

Unfermented Wine (Grape Juice) Similar to Bible Times

Recipes and Sources

1. Pharaoh's Fresh Wine (Genesis 40:11)

Get a fresh bunch of grapes. With clean hands, squeeze directly into a cup or bowl. Drink fresh and undiluted; or dilute with water to taste.

If your hands give out, place grapes in a bowl and press grapes with a potato masher.

Strain into bowl or glass.

Variation: Go to a vineyard and purchase fresh wine grapes to make your new wine.

2. Grape Molasses Wine

Purchase Pekmez, or Grape Molasses (available at larger grocery stores or on the internet)

Dip or pour small amount into small glass. Add about five parts water; stir. If a little too syrupy, add a little more water.

This is the ancient type wine that was cooked or boiled down. This thick wine was used (and still is) as preserves, sweetening, flavoring. Or it was reconstituted with water and drunk as sweet wine. Or it can be mixed with milk. Check out pekmez recipes on the internet.

This reduced wine goes by various names: Grape Molasses, Pekmez, Saba, Sapa, Dibs, Vin Cotto. My Grape Molasses was produced in Lebanon and purchased through amazon.com.

3. Fresh Filtered Un-pasteurized Grape Juice (Wine)

Draper Valley Vineyard, 1751 Draper Valley Rd., Selma, Oregon 97538.541/597-4737; drapervalleyvineyard.com

Their grape juice tastes like sweet wine, not alcoholic wine. The flavor changes over time and gets slightly sweeter with age. Draper Valley Vineyard takes wine grapes, presses them, filters and bottles it. Their process of filtration has been determined to be as safe or safer than pasteurization. All their products are nonalcoholic. Varieties include Pinot Noir (my favorite), Chardonnay, Riesling, Cabernet Sauvignon, Early Muscat, Gewurtztraminer. These are wine grape varieties made into fresh grape juice rather than allowed to rot and ferment into a dangerous drug.

4. Other Premium Grape Juice (Sweet Unfermented Wine).

Sweetwater Cellars, LLC, PO Box 1164, Clackamas, OR 97015-1164. 503/799-3003; www.sweetwatercellars.com. Sells all types of nonalcoholic gourmet grape juice from vineyards around the world. Also apple cider (a form of shekar).

Note: Some grape juice is made to taste like alcoholic wine; nondrinkers will probably not like this type grape juice. Ask if it tastes like alcoholic wine, or sweet grape juice. Of course be sure it is itself, nonalcoholic. And remember that personal tastes differ.

5. Welches Grape Juice

And then there are always the old standbys, Welches Grape Juice, Ocean Spray, Langer's, etc. Light varieties with less sugar and calories are also available.

6. Fresh Grapes and Raisins

Or just eat the grapes fresh or in the form of raisins. This was also a popular way of consuming grapes in ancient times. Spend time looking; you will find available a wide variety of raisins. See chapter 2 on recipes for unfermented raisin wine. Also see Jewish cookbooks.

7. Lactic Fermentation

See chapter 2, and Bibliography, on lactic fermentation.

8. Shekar

Shekar refers to any juice, fermented or unfermented, from fruit other than grapes. One example of nonalcoholic shekar would be un-pasteurized Bragg Apple Cider Vinegar. Other examples would be nonalcoholic cider from such places as Louisburg Cider Mill, Louisburg, Kansas (louisburgcidermill.com); and Sweetwater Cellars mentioned above.

Other Books By The Brumbelows:

The Wit and Wisdom of Pastor Joe Brumbelow: Favorite illustrations, personal stories, humor, history, folklore, and lessons learned from over 50 years in the ministry, by his son David R. Brumbelow, 240 pages; $14.95

"Some of the funniest stories you will ever read." -**John Hatch**

"A beautiful book, easy to read, and full of genuine spiritual wisdom." -**Adrian Rogers**

"Already it has been a source of illustrations for me." -**Paige Patterson**,

Masterpieces From Our Kitchen, Cookbook by Mrs. Joe (Bonnie) Brumbelow, $10

233 recipes all personally used and endorsed by a pastor's wife and mother of three boys. Most are easy to prepare; includes Mexican, Chinese, and Sugar-Free recipes, many original. Includes tamale, and egg roll recipes. Both books make great gifts.

Order signed copies from: **David Brumbelow, P.O. Box 300, Lake Jackson, TX 77566 USA**. Prices are postpaid.

Books by Free Church Press:

A Gentle Zephyr - A Mighty Wind by J. Gerald Harris; $14.95 + shipping

Urgent: Igniting a Passion for Jesus by Joe Donahue; $11.95 + shipping

Ancient Wine and the Bible: The Case for Abstinence by David R. Brumbelow; $21.00 + shipping

Free Church Press, P.O. Box 1075, Carrollton, Georgia 30112 USA; freechurchpress.com

Author's Notes

1 International Standard Bible Encyclopedia, James Orr, General Editor, Eerdmans, 1955, vol. II.

2 Pliny. XIV, Loeb Classical Library; c. AD 70.

3 Athenaeus, c. AD 230.

4 Pliny. XIV, Loeb Classical Library; c. AD 70.

5 Athenaeus: the Deipnosophists - Book 2

6 Multiple Sermons From Dr. Adrian Rogers; lwf.org.

7 Pliny. XIV, Loeb Classical Library; c. AD 70.

8 Pliny, NH, Book 23.19; c. AD 70.

9 Vitruvius Pollio, The Ten Books on Architecture, 8.3; c. 20 BC.

10 Prof. Dr. MetinSaipSürücüo□lu, Grape/Fruit Molasses, Research Faculty Lale S. Çelik. turkish-cuisine.org; accessed August, 2010.

11 Aristotle, Meteorology, Book IV; c. 350 BC., vol. I; The Complete Works of Aristotle, Princeton University Press, Princeton, New Jersey; 1984

12 Hippocrates, On Regimen in Acute Diseases, c. 400 BC; translated by Francis Adams.

13 Plutarch (c. AD 120), Why New Wine Doth Not Inebriate As Soon As Other.

14 Pliny, NH, Book 23.22; c. AD 70.

15 Athenaaeus: the Deipnosophists – Book 2

16 Plato, Letters; c. 340 BC.

17 Dr. Nelson L. Price, Pastor Emeritus, Roswell Street Baptist Church, Georgia. SBC leader and author. nelsonprice.com

18 Pliny, NH 14.6; AD 70.

19 Andrew Dalby, *Food in the Ancient World From A to Z*, Routledge, London & New York; 2003. This books is an encyclopedia

of food in classical times. Dalby is a historian, linguist, and author of several books on the ancient world.

20 Ibid.

21 Ibid.

22 Ibid.

23 Dr. Robert P. Teachout, Wine; 1983. Teachout wrote his Dallas Theological Seminary doctoral dissertation on *The Use of Wine in the Old Testament.*

24 A composite of what I've heard and read numerous times by educated men in the pulpit, classroom, books, and the internet.

25 Direct quote from a longtime professor with an earned doctorate; an otherwise very good Bible expositor. No name given to avoid singling out people and avoid embarrassment.

26 Direct quote from an educated pastor. Again, typical of modern day arguments.

27 Thomas Pellechia, Wine: The 8,000 Year-Old Story of the Wine Trade, Running Press, Philadelphia, London; 2006.

28 Andrew Dalby, Food in the Ancient World From A to Z, Routledge, London & New York; 2003.

29 From a winemaking website.

30 Andrew Dalby, Food in the Ancient Word From A to Z, Routledge, London & New York; 2003.

31 Columella, Book XII, Loeb Classical Library; c. AD 60.

32 Columella, Book XI, Loeb Classical Library; c. AD 60.

33 Varro, On Agriculture; c. 36 BC.

34 Cato, On Agriculture; c. 270 BC.

35 Aristotle, Meteorology; c. 350 BC. Vol. I.

36 Virgil (or Vergil; or Vergilius), Georgics 1; c. 29 BC. Virgil (70-19 BC) was a Roman poet and author of Bucolics, Georgics, Aeneid.

37 Virgil, The Georgics IV; c. 29 BC.

38 Pliny, NH, 14.11

39 Dr. Jim Richards, Executive Director, SBTC, Southern Baptist Texan, July 31, 2006.

40 Andrew Dalby, Food in the Ancient World From A to Z, Routledge, London & New York; 2003.

41 A Latin Dictionary, by Charlton T. Lewis, Ph.D. and. Charles Short, LL.D. Oxford. Clarendon Press. 1879.

42 Tishendorf, Travels in the East, London; 1847. Quoted by Field.

43 Dr. Nelson L. Price, pastor in Georgia and Louisiana, SBC leader and graduate of New Orleans Baptist Theological Seminary, Mercer University.

44 Patrick E. McGovern, Ancient Wine: The Search For The Origins Of Viniculture, Princeton University Press, Princeton, New Jersey, 2003.

45 Richard Teachout, On the Fruit of the Vine, 2010.

46 Annie Proulx, Making the Best Apple Cider, Storey Publishing, MA; 1980.

47 New York Times, 12-22-1999; Dining and Wine.

48 gourmed.com

49 giverecipe.com, a website giving recipe for tahini

50 turkishcookbook.com; 2007.

51 Prof. Dr. MetinSaipSürücüo☐lu, Grape/Fruit Molasses, Research Faculty Lale S. Çelik. turkish-cuisine.org/english; accessed August, 2010.

52 Pliny, Book XIV; c. AD 70.

53 Ibid

54 Varro, On Agriculture, Loeb Classical Library; c. 36 BC

55 Andrew Dalby, Food in the Ancient World From A to Z,

Routledge, London & New York; 2003.

56 Moses, c. 1450 BC; Genesis 40:11. Moses was the great leader of Israel and writer of the first five books of the Bible.

57 New York Times, To Your Health, by Florence Fabricant; Published: April 26, 1992.

58 William Patton, Bible Wines; 1871.

59 Leon C. Field, Oinos; 1883. Field was a Methodist scholar.

60 Antiquities of the Christian Church, vx 2,3; quoted by Field.

61 Acts and Martyrdom of Matthew , Sec. 25, from the second and third centuries AD; quoted by Field.

62 Jewish Encyclopedia; 1906.

63 Frederic Richard Lees & Dawson Burns, Temperance Bible Commentary; 1870.

64 Information gleaned from the magazine, *Fruit Gardener, California Rare Fruit Growers*, July/August, 1999; article by Lon J. Rombough.

65 Leon C. Field, Oinos; 1883.

66 Pliny, Loeb Classical Library, Book XIV; c. AD 70.

67 Andrew Dalby, *Food in the Ancient World From A to Z*, Routledge, London & New York; 2003.

68 Aristophanes, The Seasons; c. 400 BC. Quoted by Dalby.

69 Varro, On Agriculture, Loeb Classical Library; c. 36 BC

70 Alphonse Loubat, American Vine Dresser's Guide, G. & C. Carwell, NY; 1827.

71 New Family Receipt Book, London; 1820.

72 Author Thomas Meehan, The Gardener's Monthly And Horticulturist V21, Publisher Charles H. Marot; 1879.

73 Mike & Nancy Bubel, Root Cellaring, Storey Publishing; 1991.

74 from turkish-cusine.org 2010

75 Polybius, IV, 6, 2; c. 100 BC, Greek Historian.

76 International Standard Bible Encyclopedia, William B. Eerdmans, Grand Rapids, Michigan; 1988.

77 OK Kosher Certification article online, okkosher.com; 2002.

78 Pliny, NH, 14.11

79 Pliny. XIV, Loeb Classical Library; c. AD 70.

80 Pliny, NH, Book 23.18; c. AD 70.

81 from Horace; Pliny.

82 A. Cornelius Celsus, On Medicine, vol. II, Book IV; c. AD 50.

83 William Smith, LLD, Ed., Dictionary of Greek and Roman Geography; 1854.

84 Patrick E. McGovern, *Ancient Wine*, Princeton University Press; 2003. McGovern is one of the foremost authorities on wine, ancient and modern. He is a Senior Research Scientist in the Museum Applied Science Center for Archaeology (MASCA) and is Adjunct Associate Professor of Anthropology at the University of Pennsylvania.

85 Ibid.

86 Bersalibi, quoted by G. W. Samson.

87 quoted by Leon C. Field, Oinos, 1883.

88 Ibid.

89 Joan Nathan, Jewish Cooking in America; 1994. Mordecai Noah (1785-1851) was the first Jew in America to rise to national prominence. He was a politician, diplomat, journalist.

90 Quoted by Dr. Stephen M. Reynolds, The Biblical Approach to Alcohol; 2003.

91 The Divine Law As To Wines by Dr. G. W. Samson; 1880.

92 261, De Re Coquinaria, of Apicius; Book VII, Sumptuous Dishes; c. AD 400. Apicius is the name of a collection of Roman cooking recipes.

93   Florence Greenberg, Jewish Cookery; 1947. Greenberg cooked for a family of 12 and wrote cookery articles for the Jewish Chronicle.

94   Joan Nathan, Jewish Cooking in America; 1994.

95   Mordecai M. Noah, quoted by Joan Nathan, Jewish Cooking in America; 1994.

96   Cato, On Agriculture, Loeb Classical Library; c. 170 BC.

97   Columella, Book XII, Loeb Classical Library; c. AD 60.

98   Pliny, NH, 14.11; c. AD 70.

99   For example, un-pasteurized Bragg Apple Cider Vinegar, with the "Mother." Bragg Live Food Products, Inc., Box 7, Santa Barbara, CA 93102 USA. This would have been the type vinegar recommended by ancients such as Hippocrates; most modern day vinegar is pasteurized thereby killing any live culture.

100   Many "lactic fermentation" recipes to preserve food are given by ancient writers. For space considerations, they are not given in this book.

101   Beatrice Trum Hunter, Fermented Foods and Beverages, An Old Tradition, Keats Publishing, New Canaan, Connecticut; 1973.

102   Preserving Food Without Freezing or Canning: Traditional Techniques Using Salt, Oil, Sugar, Alcohol, Vinegar, Drying, Cold Storage, and Lactic Fermentation, by the Gardeners & Farmers of Terre Vivante, Chelsea Green Publishing Company, White River Junction, Vermont; 2006.

103   Cato, On Agriculture, Loeb Classical Library; c. 175 BC.

104   Ibid. Note the lead boiler in this quote; the Romans and ancients had no knowledge of lead poisoning.

105   Ibid.

106   Pliny. XIV, Loeb Classical Library; c. AD 70.

107 Ibid.

108 Ibid.

109 Columella, Book XII, Loeb Classical Library; c. AD 70.

110 Athenaeus, in *Deipnosophistae*, or *The Sophists at Dinner;* c. AD 220.

111 Sally Fallon, Nourishing Traditions, New Trends Publishing, Washington, DC; 2001.

112 Pliny. XIV, Loeb Classical Library; c. AD 70.

113 See ancient quotes on filtering wine in chapter 3.

114 Suma Theologica

115 Aristotle; Meteorology, Book IV; c. 350 BC., vol. I; The Complete Works of Aristotle, Princeton University Press, Princeton, New Jersey; 1984.

116 Ibid.

117 Aristotle, Princeton edition, translated by E. W. Webster.

118 Aristotle, Meteorology; c. 350 BC. Vol. I.

119 Aristotle, Meteorology, Book IV.

120 Aristotle; Poetics; c. 350 BC.

121 Aristotle; Meteorology; c. 350 BC. Vol. I.

122 Aristotle, Topics, Book II, Part 3; c. 350 BC.

123 Aristotle, Eudemian Ethics.

124 Aristotle, Poetics; c. 350 BC.

125 Hippocrates, On Regimen in Acute Diseases, c. 400 BC; translated by Francis Adams.

126 Homer, The Iliad; c. 800 BC.

127 Athenaeus, in Deipnosophistae, or The Sophists at Dinner; c. AD 230.

128 Ibid.

129 Ibid.

130 Ibid.

131 Ibid.

132 Ibid.

133 Vitruvius Pollio, The Ten Books on Architecture; c. 15 BC.

134 Athenaeus, in Deipnosophistae, or The Sophists at Dinner; c. AD 220.

135 Ibid.

136 Ibid.

137 Ibid.

138 Ibid.

139 Athenaeus: the Deipnosophists - Book 2.

140 Ibid.

141 Ibid.

141 Athenaeus, Book VI, sect. 89, Voyage of Nymphodorus, the Syracusan.

142 Columella, Book XII, Loeb Classical Library; c. AD 70.

143 Ibid.

144 Ibid.

145 Ibid.

146 Ibid.

147 Ibid.

148 Ibid.

149 Ibid.

150 Ibid.

151 Ibid.

152 Columella, De Re Rustica, vol. I.

153 Pliny, Book XII; c. AD 70.

154 Pliny, Book XIV, Loeb Classical Library; c. AD 70.

155 Ibid.

156 Ibid.

157 Ibid.

158 Ibid.

159 Ibid.

160 Ibid.

161 Ibid.

162 Ibid.

163 Ibid.

164 Footnote to Pliny's comment: in B. xxx. c. 15.

165 Pliny, Natural History, Book 9.82; c. AD 70.

166 Pliny. XIV, Loeb Classical Library; c. AD 70.

167 Pliny, NH, Book 23.20; c. AD 70.

168 Pliny, NH, Book 23.22; c. AD 70.

169 Pliny. XIV, Loeb Classical Library; c. AD 70.

170 Zenophon, The Anabasis, in the retreat of the 10,000; c. 300 BC.

171 Varro, On Agriculture, Loeb Classical Library; c. 36 BC.

172 Ibid.

173 Plutarch, Why New Wine Doth Not Inebriate As Soon As Other; c. 120.

174 Ibid.

175 Plutarch, Whether Wine Ought to Be Strained or Not; c. AD 120.

176 Andrew Dalby, Food in the Ancient World From A to Z; 2003.

177 Galen, On the Natural Faculties; c. AD 190.

178 Virgil, quoted by Dr. G. W. Samson.

179 Theophrastus, Inquiry Into Plants, Concerning Odors; c. 280 BC.

180 Theophrastus, De CausisPlantarum; c. 280 BC.

181 Ibid.

182 Ibid.

183 Galen, On Simple Medicaments, iv. 12 (vol. xi, p. 656. Quoted in -Theophrastus, De CausisPlantarum; c. 285 BC.

184 Flavius Philostratus, The Life of Apollenius; Translated by F.C. Conybeare; c. AD 240.

185 Homer, The Iliad, 12.307-330; tr. Ian Johnston; c. 800 BC.

186 Quoted by Dr. G. W. Samson. The Talmud is composed of Jewish laws and writings, c. AD 200-500.

187 Aretaeus, De curationeacutorummorborumlibri duo, Aret. CA 1.2, c. AD 60. Francis Adams LL.D., Ed.

188 Strabo, 6.1, Geography; c. AD 20.

189 Nicander of Colophon, Georgica, Fragment 86; c. 150 BC. Quoted by Robert P. Teachout, *The Use of Wine in the Old Testament*. Doctoral Dissertation at Dallas Theological Seminary, Dallas, TX; 1979.

190 Anacreon, Ode 51, c. 500 BC. Quoted in Robert Teachout's Dissertation.

191 Patrick E. McGovern, Ancient Wine; 2003. From an ancient Hittite tablet, quoted by McGovern.

192 Flavius Josephus, Antiquities of the Jews, c. AD 90, William Whiston, A.M., Ed.

193 Ibid.

194 Plato, Letters; c. 340 BC.

195 *The Living Webster Encyclopedic Dictionary of the English Language*, Delair Publishing; 1980.

196 William Smith, LLD, William Wayte, G. E. Marindin, Ed.

Article on DIONY'SIA, A Dictionary of Greek and Roman Antiquities; 1890.

197 Dr. Robert P. Teachout, The Use of Wine in the Old Testament: Doctoral Dissertation, Dallas Theological Seminary; 1979. Parenthesis are Teachout's.

198 Andrew Dalby, Food in the Ancient World From A to Z, Routledge, London & New York; 2003.

199 Ibid.

200 Ibid.

201 Ibid.

202 Dr. G. W. Samson, The Divine Law As To Wines, New York; 1880. George Whitefield Samson was a graduate of Brown University and Newton Theological Seminary. He was a highly educated, respected Baptist pastor and president of George Washington University and later at Rutgers.

203 Dr. Lyman Abbott, A Dictionary of Religious Knowledge, Harper & Brothers, New York; 1876.

204 Dr. Robert P. Teachout, The Use of Wine in the Old Testament: Doctoral Dissertation, Dallas Theological Seminary; 1979.

205 Jean Bottero, The Oldest Cuisine in the World: Cooking in Mesopotamia, translated by Teresa Lavender Fagan, University of Chicago Press; 2004.

206 Information from several sources including McGovern.

207 Patrick E. McGovern, Ancient Wine; 2003.

208 Joan P. Alcock, Food in the Ancient World; 2006.

209 Professor Murphy, D.D., Belfast, Commentary. Quoted in the Temperance Bible Commentary.

210 Dr. John Kitto, Popular Cyclopedia of Biblical Literature, Fredonia Books, Amsterdam, The Netherlands; 1852.

211 Dr. Adam Clark, quoted in Communion Wine and Bible Tem-

perance by William M. Thayer, 1869.

212 Aristotle, Poetics; c. 350 BC.

213 Dr. Jim Richards, Executive Director, SBTC, Southern Baptist Texan, July 31, 2006.

214 Ibid.

215 Dr. Jerry Sutton, SBC Pastor's Conference.

216 Dr. Stephen Reynolds, The Biblical Approach to Alcohol, L. L. Reynolds Foundation, Glenside, PA; 2003.

217 Dr. Kenneth O. Gangel, Holman New Testament Commentary, editor Dr. Max Anders, Broadman& Holman, Nashville; 2000.

218 Dr. Robert Young, Young's Analytical Concordance to the Bible, Eerdmans, 1970.

219 Yael Zisling, Winemaking in Israel - A Modern Industry Based on Ancient Traditions. Published by gemsinisrael.com; Spotlighting Israel's Lesser Known Tourist Attractions and Travel Sites, the Gems, August / September, 2001.

220 Ibid.

221 Jewish Encyclopedia; 1906.

222 Encyclopaedia Judaica; 1971.

223 Dr. Robert Young, Young's Analytical Concordance to the Bible, Eerdmans, 1970.

224 Dr. Robert P. Teachout, The Use of Wine in the Old Testament: Doctoral Dissertation, Dallas Theological Seminary; 1979.

225 Ibid.

226 *The Living Webster Encyclopedic Dictionary of the English Language*, Delair Publishing; 1980.

227 Dr. Robert Young, Young's Analytical Concordance to the Bible, Eerdmans, 1970.

228 Dr. G. W. Samson, The Divine Law As To Wines, New York; 1880.

229 Dr. Stephen Reynolds, The Biblical Approach to Alcohol, L. L. Reynolds Foundation, Glenside, PA; 2003.

230 Jewish Encyclopedia; 1906.

231 Quoted by Dr. Robert P. Teachout, The Use of Wine in the Old Testament: Doctoral Dissertation, Dallas Theological Seminary; 1979.

232 Dr. Lyman Abbott, A Dictionary of Religious Knowledge, Harper & Brothers, New York; 1876.

233 John W. Haley, Alleged Discrepancies of the Bible; 1874.

234 Dr. Robert Stein, Christianity Today magazine, June 20, 1975.

235 Dr. John A. Broadus, Matthew; Kregal: Grand Rapids; 1990.

236 Dr. Charles Wesley Ewing, The Bible and its Wines; 1985.

237 Ferrar Fenton of England, The Bible and Wine.

238 Dr. Moses Stuart (AD 1780-1852), in a letter to Dr. Nott. Quoted in Field.

239 Dr. Moses Stuart (1780-1852).

240 Dr. R. L. Sumner, Fights I Didn't Start, and Some I Did, Biblical Evangelism; 2009.

241 Dr. John R. Rice, The Double Curse of Booze, Sword of the Lord, Murfreesboro, TN; 1960.

242 Athenaeus, c. AD 220; quoted by Dr. G. W. Samson.

243 Athenaeus, Book VI, sect. 89, Voyage of Nymphodorus, the Syracusan; 320 BC.

244 Plato, Letters; c. 340 BC.

245 Dr. Lees and Dawson, Temperance Bible Commentary.

246 Aristotle; Meteorology, Book IV; c. 350 BC., vol. I; The Complete Works of Aristotle, Princeton University Press, Princ-

eton, New Jersey; 1984.

247 Patrick E. McGovern, Ancient Wine; 2003.

248 Andrew Dalby, Food in the Ancient World.

249 Dr. Jim Richards, Southern Baptist Texan, July 31, 2006.

250 Quoted in Temperance Bible Commentary.

251 Pliny, Book XII; c. AD 70.

252 Theophrastus, De CausisPlantarum; c. 280 BC.

253 Hippocrates, Aphorisms; c. 400 BC.

254 Dr. Robert P. Teachout, The Use of Wine in the Old Testament: Doctoral Dissertation, Dallas Theological Seminary; 1979.

255 Moses Stuart in a letter to Dr. Eliphalet Nott; quoted from Oinos by Leon C. Field.

256 Albert Barnes, Barnes' Notes on the New Testament.

257 Ferrar Fenton of England, The Bible and Wine.

258 Ibid.

259 Cannon Farrar; quoted in Smith's Dictionary.

260 Quoted in Wycliffe Bible Encyclopedia, Editors: Charles F. Pfeiffer, Howard F. Voss, John Rea. Moody Press, Chicago; 1975.

261 Adrian Rogers, The Battle of the Bottle. He was pastor of Bellevue Baptist Church, Memphis, Tennessee, elected three times as president of the SBC.

262 Dr. Jack Graham, BP, September 13, 2005. Graham is pastor of Prestonwood Baptist Church, Plano, TX, and past SBC president.

263 B. N. Defoe, A Complete English Dictionary, London; 1735.

264 Encyclopedia Americana, Boston; 1855. Quoted by Thayer.

265 Quoted by William M. Thayer, Communion Wine; 1869.

266 *The Living Webster Encyclopedic Dictionary of the English Language*, Delair Publishing; 1980.

267 Information from business.timesonline.co.uk.

268 Dr. Mian N. Riaz, Ph.D., presently working as a food scientist with Texas A&M University.

269 Dr. Jim Richards, Southern Baptist Texan, July 31, 2006.

270 Dr. Barry Creamer, Criswell College; blog comment, God. Real. Right.

271 Dr. Charles Wesley Ewing, The Bible and Its Wines.

272 Adrian Rogers, The Battle of the Bottle - part 2, 2003; lwf. org.

273 Believer's Study Bible, Thomas Nelson, Nashville, TN; 1991 (edited by Dr. W. A. Criswell & Dr. Paige Patterson; aka Criswell Study Bible; Baptist Study Edition).

274 W. E. Vine, Vines's Expository Dictionary of Old & New Testament Words, Nelson; 1997.

275 Ibid.

276 John MacArthur, *1 Timothy*, Moody Press, 1995. Comment on 1 Timothy 3:2.

277 Kittel's Theological Dictionary of New Testament, quoted by Dr. Brad Reynolds, Vice President, Academic Services, Truett-McConnell College (from 2006 blog article).

278 Ernest Gordon, *Christ, The Apostles And Wine*, Sunday School Times; 1944.

279 Judge H. Paul Pressler, Justice for the 14th Court of Appeals, Houston, TX. Pressler was a leader in the SBC Conservative Resurgence and author of, *A Hill On Which To Die*, B&H; 1999. W. A. Criswell said of Pressler, "A layman as devoted to Christ and the Bible as any clergyman who ever lived." Quote from Pressler's personal endorsement of Ancient Wine and the Bible.

280 Dr. Paige Patterson; president SWBTS; Baptist Press

(bpnews.net), July 7, 2006.

281 Dr. Jim Richards, Southern Baptist Texan, July 31, 2006.

282 Orin B. Whitmore, Bible Wines vs. the Saloon Keeper's Bible, Press of the Alaska Printing Co., Seattle; 1911.

283 Ibid.

284 Leon C. Field, Oinos.

285 Dr. Paige Patterson; president SWBTS; Baptist Press (bpnews.net), July 7, 2006.

286 Ibid.

287 Chrysostom (c. AD 347-407), Homily 22 on John.

288 Augustine (AD 354-430), one of the most influential Church Fathers.

289 H. A. Ironside, John, Loizeaux Brothers, Neptune, New Jersey; 1942, 1978.

290 F. B. Meyer, Gospel of John.

291 Dr. F. F. Bruce, The Gospel of John, Eerdmans; 1983.

292 Dr. G. W. Samson, The Divine Law As To Wines; 1880.

293 Adam Clarke Commentary.

294 Dr. Daniel D. Whedon (1808-1885), was a Methodist university professor, theologian, and author. Among other disciplines, he served as Professor of Ancient Languages. Quoted by Field.

295 Albert Barnes, Barnes' Notes, 1884-85 edition, reprinted by Baker Book House, Grand Rapids, Michigan.

296 Ibid.

297 Dr. Jim Anderson, Th.D., A Biblical Study on Wine, 1980, Belton, MO. Anderson is a professor at Midwestern Baptist Theological Seminary.

299 Albert Barnes, Barnes' Notes, 1884-85 edition, reprinted by Baker Book House, Grand Rapids, Michigan.

300 Dr. J. Vernon McGee, Thru The Bible, Proverbs 20 note, Thomas Nelson, Nashville, TN; 1982.

301 Dr. John R. Rice, The Double Curse of Booze, Sword of the Lord, Murfreesboro, TN; 1960. Rice (1895-1980) was educated at Decatur Baptist College (now DBU), Baylor University, University of Chicago, and Southwestern Baptist Theological Seminary. He was a prolific author, founder and editor of *Sword of the Lord*, and an independent Baptist Evangelist.

302 Dr. Robert P. Teachout, Wine; 1983.

303 O. O. Irvin, personal conversation a few years ago.

304 Dr. Paige Patterson, SWBTS, BP, July 7, 2006.

305 Dr. Jim Richards, Southern Baptist Texan, July 31, 2006.

306 Hendiadys - a figure in which a single idea is expressed by two words connected by a conjunction.

307 Dr. Robert Young, Young's Analytical Concordance to the Bible, Eerdmans, 1970.

308 William Gesenius, A Hebrew and English Lexicon of the Old Testament Including the Biblical Chaldee, 1865. Quoted in Teachout.

309 Leon C. Field, Oinos: A Discussion of the Bible Wine Question, Phillips & Hunt, New York; 1883.

310 Jerome, Letter to Nepotian, c. AD 400.

311 Dr. Lyman Abbott, A Dictionary of Religious Knowledge, Harper & Brothers, New York; 1876.

312 Dr. Stephen M. Reynolds; The Biblical Approach to Alcohol, L. L. Reynolds Foundation, Glenside, PA; 2003.

313 Dr. Frederic Richard Lees, F.S.A. & Rev. Dawson Burns, M.A., Temperance Bible Commentary, New York; 1870.

314 Dr. Robert P. Teachout, Wine, The Biblical Imperative: Total Abstinence; 1983.

315 Dr. John Kitto, D.D., S.F.A., Kitto's Cyclopedia. Quoted in

Bible Wines, William Patton.

316 Dr. John Kitto, Popular Cyclopedia of Biblical Literature, Fredonia Books, Amsterdam, The Netherlands; 1852.

317 Dr. F. R. Lees & Dawson Burns, The Temperance Bible Commentary.

318 John W. Haley, *Alleged Discrepancies of the Bible*; 1874.

319 Wycliffe Bible, John Wycliffe; c. AD 1395.

320 Patrick E. McGovern, Ancient Wine; 2003.

321 Dr. Samuele Bacchiocchi, Wine in the Bible; 2004.

322 Dr. Moses Stuart, Letter to Rev. Dr. Nott, February 1, 1848. Quoted by Patton.

323 Dr. Brad Reynolds, Vice President, Academic Services, Truett-McConnell College (from article, Alcohol Abstinence: Bias or Biblical?; 7-18-2006).

324 Andrew Dalby, Food in the Ancient World From A to Z, Routledge, London and New York; 2003.

325 Polybius, Histories, The Lotus; c. 150 BC.

326 HL Rory McGowen, Non Alcoholic Beverages of the Middle Ages; forgottonsea.org.

327 Adapted from The 'Libre de Diversis Medicinis' in the Thornton Manuscript (MS. Lincoln Cathedral, A.5.2)_. Edited by Margaret Sinclair Ogden. Published for the Early English Text Society by Humphrey Milford, Oxford University Press. Amen House, E.C. 4. England. 1938. Text c. AD 1400. (forgottensea.org).

328 Prof. Dr. MahmutTezcan, Teaching Staff, Ankara University, Faculty of Educational Sciences, turkish-cusine.org 2009.

329 Pliny. XIV, Loeb Classical Library; c. AD 70.

330 Herodotus, The Histories; c. 450 BC.

331 Xenophon, Anabasis, 1.5; c. 370 BC.

332 Flavius Josephus, The Wars of the Jews; c. AD 70. William Whiston, A.M., Ed.

333 Peter Masters, Should Christians Drink?, The Wakeman Trust, London, 1992, 2001. Note on 1 Timothy 5:23.

334 Ernest Gordon, Christ, The Apostles And Wine, Sunday School Times; 1944.

335 Paige Patterson; president SWBTS; Baptist Press (bpnews. net), July 7, 2006.

336 Pliny, Natural History; c. AD 70. Also quoted by Samson.

337 G. W. Samson, The Divine Law As To Wines.

338 Athenaeus: the Deipnosophists - Book 2; c. AD 220.

339 Mayo Clinic, accessed 4-2010. http://www.mayoclinic.com/health/food-and-nutrition/AN00576

340 Kenneth Stanley is a mechanic, motor machinist, and deacon in Northside Baptist Church, Highlands, TX; 2009.

341 Gleaned from Welch's Grape Juice website; 2009, www.welchs.com.

342 September 15, 2006 (Fisher Center for Alzheimer's Research Foundation); alzinfo.org.

343 Basil, Bishop of Cappadocia, Asia Minor; c. AD 379. Quoted in The Divine Law As To Wines by Dr. G. W. Samson.

344 Law Book of the Ante-Nicene Church, Section 54; c. AD 300. Quoted by Lees.

345 Dr. Frederic Richard Lees, F.S.A. & Rev. Dawson Burns, M.A., Temperance Bible Commentary, New York.

346 Dr. Paige Patterson, president, Southwestern Baptist Theological Seminary (swbts.edu). Baptist Press (bpnews.net), July 7, 2006.

347 B. H. Carroll (AD 1843-1914), *Interpretation of the English Bible*. Carroll was pastor of First Baptist Church, Waco, TX and founding president of Southwestern Baptist Theological

Seminary.

348 John MacArthur, *1 Timothy*, Moody Press; 1995.

349 Paige Patterson, *Living in the Hope of Eternal Life* (Titus), Wipf & Stock Publishers, 1968, 2007.

350 G. W. Samson, The Divine Law As To Wines.

351 Dr. Jim Anderson, Th.D., A Biblical Study on Wine, 1980, Belton, MO.

352 Ernest Gordon, Christ, The Apostles And Wine, Sunday School Times; 1944.

353 Orin B. Whitmore, Bible Wines vs. the Saloon Keeper's Bible, Press of the Alaska Printing Co., Seattle; 1911.

354 Dr. Paige Patterson; president SWBTS; Baptist Press (bpnews.net), July 7, 2006.

355 A. C. Dixon; 1891, quoted by Coker. A. C. Dixon (1854-1925) was a pastor in North Carolina, Maryland, and New York. He also pastored Moody Memorial Church, Chicago, Metropolitan Tabernacle, London, and worked with R. A. Torrey on the set of books, *The Fundamentals*.

356 Deets Picket, editor, Cyclopedia of Temperance, Prohibition, and Public Morals, Methodist Book Concern, New York, Cincinnati; 1917.

357 Dr. Stephen M. Reynolds, Biblical Approach to Alcohol; 2003.

358 Dr. Herschel H. Hobbs, The Baptist Faith & Message, Convention Press, Nashville, TN; 1971.

359 Dr. Robert P. Teachout, Wine; 1983.

360 Flavius Josephus, Antiquities of the Jews, c. AD 90, William Whiston, A.M., Ed.

361 Ernest Gordon, Christ, The Apostles And Wine, Sunday School Times; 1944.

362 Dr. G. W. Samson, The Divine Law As To Wines.

363 Cyril, c. AD 440, speaking of Acts 2:13; quoted by Dr. G. W. Samson.

364 Wycliffe Bible, c. AD 1382. Early English translation.

365 Dr. Robert P. Teachout, Wine; 1983.

366 Dr. Stephen M. Reynolds, The Biblical Approach to Alcohol; 2003.

367 Andrew Dalby, Food in the Ancient World From A to Z; 2003.

368 Jerome, Church Father and Bible translator; c. AD 400. Quoted by Dr. G. W. Samson.

369 G. W. Samson, The Divine Law As To Wines.

370 Ernest Gordon, Christ, The Apostles And Wine, Sunday School Times; 1944.

371 Ibid.

372 Dr. Robert Young, Young's Analytical Concordance.

373 Dr. Frederic Richard Lees, F.S.A. & Rev. Dawson Burns, M.A., Temperance Bible Commentary, New York.

374 Dr. Duane A. Garrett, Proverbs, The New American Commentary, Broadman, Nashville; 1993.

375 Dr. Stephen M. Reynolds, Biblical Approach to Alcohol; 2003.

376 Dr. Frederic Richard Lees, F.S.A. and Rev. Dawson Burns, M.A., Temperance Bible Commentary, New York.

377 Donald Guthrie, The New Bible Commentary: Revised, Eerdmans, 1979.

378 Dr. Stephen M. Reynolds, Biblical Approach to Alcohol.

379 prohibition.osu.edu; 2008.

380 William J. Bennett, The Devaluing of America, Simon & Schuster, New York; 1992.

381 Ibid.

382 Time Magazine, July 17, 2008.

383 aim-digest.com; 2005.

384 Dr. R. L. Sumner; Editor, Biblical Evangelist; author (from a personal email).

385 Dr. Norman L. Geisler; Veritas Evangelical Seminary; Christian apologist, author.

386 Dr. Jim Richards, Southern Baptist Texan, July 31, 2006.

387 Reported June 21, 2011 by WebMD Health News. Reviewed by Laura J. Martin, MD. The new study appeared in *Addiction*.

388 John Wesley, 1744; founder of the Methodist Church. Quoted by Ewing.

389 Dr. Charles Wesley Ewing, The Bible and Its Wines; 1985.

390 Dr. Richard Land, *Ethics and Religious Liberty Commission, SBC*, Nashville & Washington, DC; erlc.com.

391 Jack London, John Barleycorn; 1913.

392 Lieutenant General Winfield Scott, in a November 11, 1862 letter to Edward C. Delavan.

393 Alexander MacLaren, sermon, Portrait of a Drunkard, Proverbs 23:29-35.

394 Dr. Robert A. Millikan, Nobel Prize winner in physics. Quoted by Sumner, Fights I Didn't Start.

395 Stephen M. Reynolds, Biblical Approach to Alcohol.

396 Jerry Vines, *Libertinism: A Baptist and His Booze*; jerryvines.com.

397 John W. Haley, Alleged Discrepancies of the Bible; 1874.

398 Flavius Josephus, Antiquities of the Jews, c. AD 90, William Whiston, A.M., Ed.

399 Jewish Encyclopedia, 1906.

400 Dr. Charles Wesley Ewing, The Bible and Its Wines, 1985.

401 Upton Sinclair, Cup of Fury, Fleming H. Revell Company, Westwood, New Jersey, 1956, 1965.

402 Jack London, John Barleycorn, Quill Pen Classics, New York; 1913, 2008.

403 From Dr. W. A. Criswell's book on Daniel. (wacriswell.org). Criswell (1909-2002) was pastor of First Baptist Church, Dallas, TX and SBC president.

404 R. G. Lee, Sourcebook of 500 Illustrations, Zondervan, 1964. Dr. Lee (1886-1978) was pastor of Bellevue Baptist Church, Memphis, TN, author, SBC president. His most famous sermon was, "Payday Someday."

405 nethymnal.org, hymn, "Yes, I Know."

406 Reuters, reported in Canada.com, 2-11-2011.

407 Sergeant Alvin York, Sergeant York, Broadman& Holman; 1997.

408 Pastor Beau Rosser, Highlands, TX; 2010. Quotes personally confirmed by Rosser.

409 Jerome, Letter XXII. To Eustochium, Principle Works of Saint Jerome by Philip Schaff.

410 William B. Terhune, M.D., The Safe Way to Drink: How to Prevent Alcohol Problems Before They Start, New York, William Morrow & Company, Inc., 1968.

411 Allen NE, Beral V, Casabonne D, et al. Moderate alcohol intake and cancer incidence in women. Journal of the National Cancer Institute. 2009; 101: 296-305.

412 Pliny. XIV, Loeb Classical Library; c. AD 70.

413 Philo, quoted in The Divine Law As To Wines by Dr. G. W. Samson.

414 Dr. G. W. Samson, The Divine Law As To Wines; 1880. p. 204-205.

415 Ibid.

416 Clement of Alexandria, c. AD 200. Quoted in The Divine Law As To Wines by Dr. G. W. Samson; 1880. p. 200.

417 J. Wilbur Chapman, The Life & Work of Dwight Lyman Moody.

418 S. G. Hillyer, Reminiscences of the Old-Time Baptists (1902), Baptist Classics, Treasures From the Baptist Heritage, Broadman& Holman; 1996.

419 Dr. John L. Dagg (1794-1884) Southern Baptist preacher and theologian. From the Autobiography in his Manual of Theology.

420 www.godrules.net/library/spurgeon/NEW8spurgeon_ d12.htm; accessed 4-12-2010.

About the Author

David R. Brumbelow is a graduate of Sam Houston High School, Houston, TX; East Texas Baptist University, Marshall, TX (B.A.); Southwestern Baptist Theological Seminary, Fort Worth, TX (M. Div.). He serves as pastor of Northside Baptist Church, Highlands, TX and previously pastored churches in Henderson and Beeville. He has been involved in farm and ranch work, hunting, fishing, and trapping.

Brumbelow previously authored a book about his dad, *The Wit and Wisdom of Pastor Joe Brumbelow*, Hannibal Books, 2005. He has taught Bible Courses at Lee College, Baytown and San Jacinto College, Pasadena. He enjoys gardening and the outdoors and has taught *Fruit Tree Growing* Classes at Lee College and elsewhere.

David grew up in a pastor's home. He and both his brothers, Steve Brumbelow and Mark Brumbelow, serve as pastors.

Ancient Wine and the Bible: The Case for Abstinence
© *David R. Brumbelow; 2011*

www.ingramcontent.com/pod-product-compliance
Lightning Source LLC
Chambersburg PA
CBHW061003280326
41935CB00009B/809